Geospatial Information Handbook for Water Resources and Watershed Management, Volume I

Volume I of *Geospatial Information Handbook for Water Resources and Watershed Management* discusses fundamental characteristics, measurements, and analyses of water features and watersheds including lakes and reservoirs, rivers and streams, and coasts and estuaries. It presents contemporary knowledge on Geospatial Technology (GT)–supported functional analyses of water runoff, storage and balance, flooding and floodplains, water quality, soils and moisture, climate vulnerabilities, and ecosystem services.

- Captures advanced Geospatial Technologies (GTs) addressing a wide range of water issues
- Provides real-world applications and case studies using advanced spectral and spatial sensors combined with geospatially facilitated water process models
- Details applications of ArcInfo/ArcGIS, Google Earth Engine, and other systems using advanced remote sensors, including hyperspectral ER2 AVIRIS, Sentinel-1 and -2, MODIS, Landsat 7 ETM+, Landsat 8 OLI and TIPS, SAR radar, and thermal imaging
- Global in coverage with applications contributed by more than 170 authors with lifelong expertise in water sciences and engineering

This handbook is a wide-ranging and contemporary reference of advanced geospatial techniques used in numerous practical applications at the local and regional scales and is an in-depth resource for professionals and the water research community worldwide.

Geospatial Information Handbook for Water Resources and Watershed Management

Volume I

Fundamentals and Analyses

Edited by
John G. Lyon
Lynn Lyon

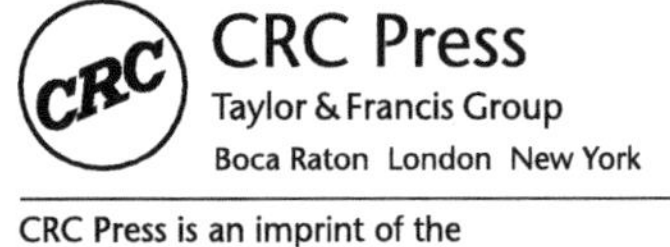
CRC Press
Taylor & Francis Group
Boca Raton London New York

CRC Press is an imprint of the
Taylor & Francis Group, an **informa** business

First edition published 2023
by CRC Press
6000 Broken Sound Parkway NW, Suite 300, Boca Raton, FL 33487-2742

and by CRC Press
4 Park Square, Milton Park, Abingdon, Oxon, OX14 4RN

CRC Press is an imprint of Taylor & Francis Group, LLC

ISBN: 978-1-032-00636-9 (hbk)
ISBN: 978-1-032-00645-1 (pbk)
ISBN: 978-1-003-17501-8 (ebk)

DOI: 10.1201/9781003175018

Typeset in Palatino
by codeMantra

Contents

Editors

Over the years, John G. and Lynn Lyon have worked on water issues with great interest and heart. A hallmark of their scholarship has been helping others with scientific authorship. Books are a great way to capture thoughts and methods and make them accessible to others worldwide. In that vein, here they have advanced a whole cadre of thought leaders with a broad focus on water.

Whether on the local, regional, continental, or global scale, water issues unite passions. Part of the challenge is the breadth and complexity of water. If many people take a piece, the whole society can fashion sustainable outcomes. This handbook is focused on presenting a number of ways to facilitate these thoughtful contributions.

John earned a PhD at the University of Michigan, and Lynn earned a master's degree at Ohio State University.

Contributors

Daniel Aja
Africa Centre of Excellence in Coastal Resilience
University of Cape Coast
Cape Coast, Ghana

Mohammed Shahidul Alam
Department of Fisheries
University of Chittagong
Chittagong, Bangladesh

Tesfaye Bayu
Department of Natural Resources Management
Debre Markos University
Burie, Ethiopia

Venkatesh Budamala
School of Civil Engineering
Vellore Institute of Technology
Vellore, India

Guo Chen
Geography, Environment, and Spatial Sciences
Michigan State University
East Lansing, Michigan, USA

Md. Zahedur Rahman Chowdhury
Institute of Marine Sciences
University of Chittagong
Chittagong, Bangladesh

Sayedur Rahman Chowdhury
Institute of Marine Sciences
University of Chittagong
Chittagong, Bangladesh

Jayantha Ediriwickrema (retired)
Durham, North Carolina, USA

Eyasu Elias
College of Natural and Computational Sciences
Center for Environmental Science
Addis Ababa University
Addis Ababa, Ethiopia

Mateo Gašparović
Faculty of Geodesy
University of Zagreb
Zagreb, Croatia

Enamul Hoque
Department of Oceanography
University of Chittagong
Chittagong, Bangladesh

Amit Jamwal
Aryabhatta Geo-Informatics and Space Application Centre
Kasumpti Shimla, India

Gordana Kaplan
Institute of Earth and Space Sciences
Eskisehir Technical University
Eskisehir, Turkey

Shyamal Karmakar
Environmental Science
Institute of Forestry and Environmental Science
University of Chittagong
Chittagong, Bangladesh

Ross Lunetta (retired)
US Environmental Protection Agency
Cary, North Carolina, USA

Junyan Luo
Center for Systems Integration and Sustainability
Michigan State University
East Lansing, Michigan, USA

John G. Lyon (retired)
US Environmental Protection Agency
Warrenton, Oregon, USA

Lynn Lyon (retired)
Warrenton, Oregon, USA

Alexander J. Macpherson
US Environmental Protection Agency
Research Triangle Park, North Carolina, USA

Amit B. Mahindrakar
School of Civil Engineering
Vellore Institute of Technology
Vellore, India

Ota Henry Obiahu
Department of Soil Science and Environmental Management
Ebonyi State University
Abakaliki, Nigeria

Vikas Kumar Rana
Water Resources Engineering and Management Institute
Faculty of Technology and Engineering
The Maharaja Sayajirao University of Baroda
Vadodara, India

Yang Shao
Department of Geography
Virginia Polytechnic Institute and State University
Blacksburg, Virginia, USA

Vikram Sharma
Faculty of Science
Department of Geography
Banaras Hindu University
Varanasi, India

B. Srimuruganandam
School of Civil Engineering
Vellore Institute of Technology
Vellore, India

Tallavajhala Maruthi Venkata Suryanarayana
Water Resources Engineering and Management Institute
Faculty of Technology and Engineering
The Maharaja Sayajirao University of Baroda
Vadodara, India

Mohammad Muslem Uddin
Department of Oceanography
University of Chittagong
Chittagong, Bangladesh

Abhinav Wadhwa
Interdisciplinary Centre for Water Research
Indian Institute of Sciences
Bengaluru, India

1

Introduction to Volume I of the Water Geospatial Handbook

John G. Lyon and Lynn Lyon

CONTENTS

Many current capabilities of geospatial analyses, modeling, and remote sensor measurements are presented. Together these are known as geospatial technologies (GT). Once the domain of geospatial experts with expensive tools, modern technologies are now widely available and accessible to a broad audience seeking to better understand a range of large-scale issues.

This volume looks at the use of GT in applications focused on water resources, land use land cover analyses (LULC), water quality, and landscape ecology arenas, among others.

History

The use of GT has allowed more complex and larger projects to be conceived and executed. Researchers typically have a given study location such as Hubbard Brook Ecosystem Study site, Agriculture Research Service sites, Forest and Range Experimental Stations, or UN Heritage Ramsar and World Heritage Convention sites, where they develop and test new techniques or

DOI: 10.1201/9781003175018-1

propose how to measure a construct with new tools. These techniques and measurements are frequently used in conjunction with legacy techniques, tools, and multiple ways of measuring to gain a broader understanding of the study area.

Traditional Approaches Augmented

In 2000, the book *GIS for Watershed and Water Resource Management* consolidated the results of some of the large GT water projects under way at the time. Many journal articles and books have been published before and since as a lot has happened in 20+ years.

Modeling, in particular, has grown exponentially. Traditional measurements coupled with data derived from GIS and remote sensing are now the standard foundation for many environmental and ecological models.

Take as an example water flows. Traditional measurements often come from gauges that provide continuous or interval-based data. Data from gauges is preferred but limited to locations where gauges already exist. New gauges are expensive to install and operate, and new or temporary sites do not have the history of old. Historic water flow data for multiple years is available from sources such as the USGS's National Water Quality Assessment or NAWQA program and their projects. Detailed coverage - but still local or regional at best.

With the addition of GT, researchers can take traditional water flow and quality measurements, and add data from remote sensors. Remote measures can provide large area coverage information. Scientists can interpolate on a scientific basis between and across the data fields. This work can be done in such a manner that facilitates scaling or interpolations based on known characteristics in the remote sensor–generated database.

Another example where geospatial information augments traditional data is in measuring land use. Land use is difficult to measure but can be inferred by using a surrogate or indicator measurement. Remote sensors can show land cover data that can be measured and quantified. Land use then can be inferred by land cover. Land cover serves as a surrogate or indicator for land use, and correlations can be made between land cover and land use (LULC). Formulas and models based on these data can be developed. Coupling processes and variable behaviors associated with land cover allows for integration across the landscape.

In large-area projects, researchers may want to look at water quality and land use, so they would use a combination of traditional water flow and quality measurements and geospatial data. Gauges provide measurements of water, and satellites provide land cover data. An entire watershed can be mapped and analyzed with this combination of traditional and geospatial data. Researchers can use these same methods with other watersheds and then compare relative water quality of one watershed to another.

Adding geospatial technologies and approaches has reduced the burdensome issue of limited data allowing researchers to get a big-picture view of larger sections of the landscape.

Building Spatial Datasets

A great advantage of remotely sensed data is the collection of uniform samples over large areas. Data is collected in a grid cell or matrix format yielding a searchable and analyzable spatial dataset. The resulting dataset can be projected as an image in two dimensions, or as a matrix cube in three dimensions. In three plus dimensions, the data can be stored as an n-dimensional hypervolume of variables or spectral values or principal component axes. The layers stacking in n-dimensions can be the multispectral or hyperspectral wavelength values, or ratio indices, or even land cover identification numbers, aka categorical variables.

Mega datasets can be fashioned with layers of variables creating a cube of data, with x and y coordinates being the grid cells navigated into a map projection and the z value being the brightness number or radiance or irradiance.

The matrix of grid cells allows one to image the materials and to operate on the image database as layers of content and foster visualization. For human interpretation, a variety of images can allow evaluation of phenomena. For machine learning or AI efforts, the variety is endless and need only be guided by theory and hypotheses. Certain procedures are valuable for data visualization and analysis, including image processing techniques, simulation modeling, virtual reality, and artificial reality.

Spatial Positioning

The advent of GPS or three-dimensional satellite positioning in X, Y, and Z dimensions, and subsequent navigation is a real advantage of GT. These GT technologies can include positioning via global positioning systems or GPS, the Russian GLONASS or the European Space Agency's Galileo global navigation satellite system or a combination of them.

Digital elevation models or DEMs can provide pixel point elevations or Z values in grid cells. The resulting products can be processed for details like slope and aspect or be used to identify low and high points. They can be used to model and route water on its pathways through gullies, streams, and rivers.

DEMs can be obtained at various mapping scales from government or commercial sources. One can make their own if elevation or Z data are available and navigated into X, Y coordinate systems. GPS, photogrammetry, or surveying can supply these details for DEM creation.

All these mapping projects can be navigated into map projections such as Universal Transverse Mercator or UTM or latitude and longitude values.

Spectral Resolution

In the current world of remote sensor imaging, a variety of spatial and spectral resolutions are available. Tools like Google Maps can help, and Google Earth Engine or GEE and other software can ingest the likes of Landsat 8, the Indian IRS and European Sentinel satellite sensors, as well as other systems processing small satellites as well as aircraft and drone-borne sensor data. Along with big data computer processing and analysis, much can be done. One still makes use of spectroscopy, spectral signature analyses, photogrammetry, surveying, geodesy, and light-ranging LiDAR and microwave sensors or Radar in these endeavors.

Advanced sensors can also measure spectral differences in materials by parsing wavelengths of light into digital range bins. This allows discrimination of the spectral characteristics of materials in relative terms and/or in absolute light energy terms upon "cleanup" of random source contributions to the spectra.

This brings detailed sensing at a distance from the sensor. Remote sensing is like laboratory spectroscopy that has been used up close in identification of chemicals and their concentrations via laboratory instruments. This is now the use of light spectra from the earth's surface to characterize those mixtures within the imaged grid cell or pixel. The mix of materials within the pixel reflects spectrally and creates a pattern. Called a spectral signature, it acts like a "fingerprint" of the pattern of materials imaged and/or their mixtures and can be subject to detailed analyses.

Many methods of spectral analysis are available to help pull detail out of the measured parameter. They can include wavelength or wave band selection (e.g., principal component analysis, band to band correlation, stepwise discriminant analysis), use of spectral indices (e.g., band ratios or normalized difference indices), linear multivariate statistics and models (e.g., multivariate regression, partial least squares regression, principal component regression), and nonlinear methods (e.g., spectral angle mapper, wavelets). These approaches are detailed in the chapters ahead.

With all these bands, it is also possible to winnow their number to only those that supply actual information for the task at hand. Feature selection is the action and methods that can reduce the array of bands to meaningful numbers while reducing the dimensionality of the dataset and "noise" associated with the multiple bands. This is particularly true of hyperspectral sensors where optimization of band selection is necessary to avoid too much duplicative or intercorrelated data.

Scaling and Modeling

Spatial computing power is used fully by models, and hence more detailed and spatially pertinent results are produced. A strength of GT is that it is possible to process the datasets using many types of numerical analysis procedures. One can process thematic and cartographically true matrix computations via linear algebra, and/or simpler methods.

The capability to store and quantify data on a spatial basis is an inherent characteristic of geospatial technologies or GT. Modeling with geospatial facilitates the approximation of processes to understand results and forcing function variables, all the while stored within a georeferenced database.

The use of statistical analyses has proven of great value in water studies over large areas.

Statistical approaches evaluate variables or phenomena as to the variability of their behavior. One can measure model accuracy and precision by traditional statistical measures such as probability and significance levels, goodness of fit via coefficients of determination, correlations, and analysis of variance and regressions with parametric or non-parametric assumptions as to distribution.

The goals are to test hypotheses and develop relational models of empirical origin. If these models are robust, the relationships can often be applied to other systems or locations and to different times of the year or season.

Deterministic Modeling

Models often take the deterministic form, where the phenomena being studied are mathematically modeled and simulated. Using numerical descriptions for the physical, chemical, or biological processes of interest results in complex models composed of submodels addressing each phenomenon or process with weighted contributions to model results.

Both statistical and deterministic models often consist of several submodel units with "fitting" coefficients. Coefficients used with this approach reflect the characteristics of natural processes, and they will adjust the contribution of variables or submodels to the overall model parameters to mimic the modeled processes.

A more "natural" coefficient better defines the behavior of model variables in mimicking processes. It also allows the modeler to achieve high fidelity between natural systems and their model simulations. To optimize the simulations of natural phenomenon and processes, the coefficients need to reflect the reality of the situation. As a given model begins to approximate nature, its further development often takes the path of improving the quality of coefficients.

Frequently, several experiments will be executed to better measure the level of a coefficient and its effect on the whole model. This to better have it mimic nature and help supply better model predictions.

Experiments using statistical or deterministic models can be greatly facilitated by the analysis of these individual model coefficients. These analyses are driven to find the sensitivity of the overall model results or simulations to a given variable.

The characterization of the behavior of a given variable and its influence on the results of model simulations is called a sensitivity analysis. Sensitivity analyses are part of good modeling strategy because it is very desirable to understand the individual contributions of model coefficients to the overall results, and to ensure that each variable and/or submodel contribution is appropriate to or like nature.

Verification

The success of a model is commonly evaluated using verification incorporating independent data sets. Verifications or accuracy assessments are part of a good modeling strategy because it is important to demonstrate precision and accuracy independent of the data sets used to "train" the model.

Remote sensor data sets are often used to fulfill the requirement for an independent data set to check model results. This can be done *a priori* by subsampling the original data set and retaining the independent subset to use at the end. Or another dataset can be obtained and used at the end. The validation can be subjected to accuracy assessment testing for correct identification of say land cover and identifying error of commission and omission. By storing the results in a two-by-two "confusion" matrix, one can judge overall accuracy and class accuracies along the matrix diagonal, and the identity of errors or confusion found in the off diagonal cells. Further analyses with Kappa statistics can help show goodness of the effort and utility.

Applications

The thoughtful management of resources can lead to the betterment of water, soils, plants, animals, other biota, and humans. Efforts over the last hundred years internationally and domestically have included forest management for watersheds and drinking water production, conservation, fire management, and land and water management. Current approaches include sustainability, landscape ecology, ecological services, and generally the thoughtful use of renewable resources. Climate solutions are being studied and implemented.

To understand the natural variability of earth processes and the impact of humans, research is necessary. Once characteristics and processes are known, one can implement approaches that parse and reduce use of resources such as water.

2

Introduction to Volume I

John G. Lyon and Lynn Lyon

CONTENTS

Select geospatial technologies and applications are presented in hopes of sparking interest. Chapter contributions were selected for their applications and technologies important to water issues. Contemporary authors were sought to feature new ideas from talented groups with integrated systems, and worldwide to make the examples appreciable and noteworthy.

These contributions focus on applications related to water and processes influenced by water. Water resources, water quality including nonpoint sources of pollution, watersheds, land cover and hydrology, ecosystem services, climate, and geospatial as well as water-land management-related applications are emphasized.

Contributions also represent advances in technologies and their applications. Technologies include advanced sensor measurements and imaging, data handling and analysis, and modeling, as well as statistical and coefficient sensitivity analyses.

The digital imaging approach to storing and processing spatial image data has been a fantastic boon to analyses. Either as a focal point or as an important adjunct, advances in geospatial-based statistical and deterministic modeling in the water arena have been valuable.

The goal, of course, is to be innovative with tools applied to the application. Therein lies the challenge and reward. We all can learn from and hopefully

DOI: 10.1201/9781003175018-2

apply them in future work. The user need only identify the utility of a given approach, and then work to adopt it within their existing suite of protocols applied to applications.

Volume I, Chapter 3, Lake Measurements and Google Earth Engine

Lakes and streams are distributed across the landscape and may have different spatial distributions depending on the regional geology, precipitation, and runoff. This application surely demonstrates the detectability of lacustrine systems via remote sensing despite their variability in size and distribution, and capabilities for monitoring water features.

"Large-scale mapping and monitoring inland waters by Google Earth Engine and remote sensing techniques" by Gordana Kaplan and Mateo Gašparović shows how multiple satellites and sensors, over an annual cycle of image gathering, can facilitate identification, and mapping of ponds, lakes, and reservoirs (Volume I, Chapter 3).

These studies were facilitated by using geospatial technologies notably by use of Google Earth Engine (GEE). A simple water index was calculated using Sentinel-2 sensor data and posed in GEE and then adjusted according to the study area. Results showed that GEE was a very successful application for handling big satellite datasets and accurately extracting water bodies on a national level.

The results can be valuable for many administrative applications where updated water maps are required and can be used in flooded areas for post-damage mapping and assessment. Results also demonstrated the utility of cloud-based computing, modeling, and storage and sharing of solutions like those provided by GEE.

Volume I, Chapter 4, Flood Mapping and Ground Water Recharge

To identify potential zones for artificial recharge of groundwater, an integrated approach was implemented using a GIS-based multi-criteria evaluation. In "Identification of potential runoff storage zones within watersheds," Vikas Kumar Rana and Tallavajhala Maruthi Venkata

Suryanarayana pursued identification of parameters that influence flooding, and mapped zones and selected suitable structures for recharge (Volume I, Chapter 4).

A Landsat-8 image was used to produce thematic maps such as geomorphology, geology, land use/land cover and lineaments. Slope and drainage networks were derived from a Terra satellite ASTER sensor Digital Elevation Model. Thematic layers on soils, rainfall, groundwater fluctuation levels, and aquifer characteristics were also prepared and weighed by a multi-criteria decision-making technique.

These thematic layers were used in the integrated model for identification of potential groundwater recharge zones in the Bhavani micro watershed. Such knowledge can be a great help in identifying and prioritizing sites for runoff storage and groundwater recharge.

Volume I, Chapter 5, Flood Risk Modeling

In "Flood risk zone mapping using a rational model in a highly weathered Nitisols of southeastern Nigeria," Daniel Aja, Eyasu Elias, and Ota Henry Obiahu modeled flow based on a Relational Rule for flood assessment (Volume I, Chapter 5). Using ArcGIS in combination with a modified rational model, the output of the model was integrated into the GIS by arithmetic overlay operation methods.

The results demonstrate that the delineated areas/sub-catchments experienced the same rainfall intensity of 414.2 mm/h, but flood extents in the areas were different. For instance, the very high flood risk zone covered about 22.8% of the study area, while the low-risk zone covers about 44.3%, and the possible areas likely to experience seasonal floods with given rainfall inputs were mostly below 40 m elevation. Results will be helpful to prioritize development efforts at the grassroots level and to formulate flood adaptation strategies considering potential climate variability in precipitation. These areas are populated by villages that can be greatly disturbed by flooding, and that knowledge can help in prioritization areas for remediation.

Volume I, Chapter 6, Flood Vulnerability Assessment

In "Morphometric indicators-based flood vulnerability assessment of upper Satluj basin, Western Himalayas, India," Amit Jamwal and Vikram Sharma studied flood incidences that were common in the upper Satluj basin of

Himachal Pradesh and as to their potential and real damages to villages and to farm lands (Volume I, Chapter 6).

Vulnerability assessments of flood scenarios were analyzed based on hydrological and geomorphological parameters. The offing of change in precipitation and runoff due to climate among others helped to drive these efforts. GIS was used to calculate the vulnerability assessment index and to populate vulnerability maps through overlay analysis. The regional vulnerability was calculated through a weighted score method.

For these indicators, the hydrological parameters with high value indicated that the basin had high vulnerability during the time of floods; this indicated that maximum losses were recorded during peak discharges which are likely going up with climate change. Several basins had the relatively high vulnerability score with high potential of flooding disasters, effectively identifying areas for study and monitoring.

Here vulnerability risk modeling and forecasting is demonstrated methodically as an informative approach to planning and management. Morphological parameters indicated that basin has high vulnerability of flood impact in terms of landslides, soil erosion, and physical degradation. The bank areas of the mainstreams need structural management and an effective management information system. The installation of such a system in the mainstream of highly vulnerable areas can reduce the impact of flood incidences.

Climate and Geospatial

Volume I, Chapter 7, Climate and Sandbars

In "Remote sensing measures of sandbars along the shoreline of Sonadia Island, Bangladesh, 1972–2006," Enamul Hoque and team characterize the nature and magnitude of oceanic deltaic sandbar's morphological changes between 1972 and 2006.

In a dynamic coast, the morphology of a sandbar is anticipated to change regularly, particularly when there is a large annual sediment supply. This is true of the deltaic plain formed by three great rivers, the Ganges, the Brahmaputra, and the Meghna.

The sandbar's changes across the shoreline were calculated using field survey data and manually drawn outlines of overlaid maps created from five different satellite images. Continuous and rapid morphological changes of the sandbar along the Island's shoreline were found. The net annual rate of shoreline displacement ranged between 3.94 and 7.79 m during the period. These likely resulted from the Island being exposed to huge wave and tidal action and excessive sediment supply around the Island. The sandbar's "head" appeared to be the most active zone with significant annual shoreline displacement of 22.00 m. The sandbar's far end was more stable with a yearly shoreline displacement of 0.50 m. Changes were more drastic between 1999 and 2006 than between 1972 and 1999.

These rapid changes after 1999 could indicate that the sandbar is becoming increasingly vulnerable to anthropogenic-related changes in precipitation and sediment loads as well as the adverse effects of sea-level rise, and climate-driven runoff of the large river system.

Volume I, Chapter 8, Climate, Agriculture, and Nitrogen Cycle Air and Water Quality

Agriculture is one of the largest contributors to greenhouse gas emissions, derived from livestock farming, with potential water quality diminishing, and increased poor quality air emissions from agricultural soils potentially due to application of excessive N fertilizers and decomposition of organic material (Volume I, Chapter 8). In "Integrated soil fertility management for climate change mitigation and agricultural sustainability," Tesfaye Bayu examines contributions of integrated fertility management to mitigate climate and sustain agricultural production.

Combined application of farmyard manure and mineral fertilizer was found to be very economical as compared to solely N and P application in maintaining sustainable agricultural productivity. From review, Bayu found that maximum sustained crop production (2.88 t/ha) was obtained when 69 kg of NP fertilizer was applied with 10 t/ha farmyard manure.

Importantly, combined application of farmyard manure and NP fertilizer contributes to agricultural sustainability. Applying integrated soil fertility increased total nitrogen and available phosphorus in the soil for agriculture. The highest level of carbon (12 mg/kg) was sequestered when farmyard manure was applied with NP fertilizer on maize- and wheat-cropped Alfisoils. Application of integrated fertility management reduces N_2O emissions by increasing nitrogen-use efficiency, and likely diminishing levels of other waterborne N compounds (Ward et al. 2015).

Volume I, Chapter 9, Climate, Agriculture, and Large Lakes

Studying issues on the Laurentian Great Lakes has been a challenge due to their size and differences in the watersheds and sub-basins characteristics.

An additional challenge is the presence of the lakes and watersheds in two countries, a province, and six states. Then there are numerous municipalities on shore and in the watershed the likes of Toronto, Chicago, Cleveland, Detroit/Windsor, Toledo, Milwaukee, Duluth/Superior, Thunder Bay, and Sault Ste. Marie.

In "Monitoring common agricultural cropping across the US and Canadian Laurentian Great Lakes Basin (GLB) watershed using MODIS-NDVI data," Ross Lunetta and team used the Moderate Resolution Imaging Spectrometer (MODIS) Normalized Difference Vegetation Index (NDVI) 16-day composite data products (MOD12Q) to develop annual cropland and crop-specific map databases on corn, soybeans, and wheat for the Great Lakes Basin (Volume I, Chapter 9).

The crop area distributions and changes in crop rotations were characterized by comparing annual crop map products for 2005, 2006, and 2007. The total acreages for corn and soybeans were relatively balanced for calendar years 2005 (31,462 and 31,283 km^2, respectively) and 2006 (30,766 and 30,972 km^2, respectively). Conversely, corn acreage increased approximately 21% from 2006 to 2007, while soybean and wheat acreage decreased approximately 9% and 21%, respectively.

Two-year crop rotational change analyses were conducted for the 2005–2006 and 2006–2007 time periods. The large increase in corn acreages for 2007 was likely due to introduced crop rotation changes across the GLB. Compared to 2005–2006, crop rotation patterns for 2006–2007 resulted in increased corn–corn, soybean–corn, and wheat–corn rotations.

The increased corn acreages could have potential negative water quality impacts on nutrient loadings, pesticide exposures, and sediment-mediated habitat degradation. Increases in US corn acreages since 2007 were likely related to biofuels mandates, while Canadian increases were attributed to higher worldwide corn prices.

You may think this is an old issue. Corn alcohol-blended gasoline is sold over the USA during winter to help gasoline burn more efficiently and reduce air emissions. The Renewable Fuel Standard is being argued again in the US Congress, and its denial of exemptions to small refineries from complying with fuel blending requirements may limit or close those refineries. Clearly, this is a tradeoff of air emissions pollution for water quality and land cover change pollution.

All of this brings in debate, regulatory authority, and the courts in questions of the National Environmental Policy Act or NEPA planning and execution. Where the question of multiple effects of forcing function variables is the point and whether multiple variables and their "Cumulative Effects" have been accounted for in Environmental Assessments or Environmental Impact Statements.

Additional long-term study is needed to determine the potential impacts of increased and continued corn-based ethanol agricultural production on watershed ecosystems and receiving waters. Another question is the tradeoff of water quality pollution versus air emissions from autos and trucks using ethanol-augmented gasoline.

The following contribution demonstrated how such a database could be used for modeling and views on changing scenarios of sedimentation in receiving watersheds.

Volume I, Chapter 10, Climate, Great Lakes, and SWAT Modeling

In "SWAT modeling of sediment yields for selected watersheds in the Laurentian Great Lakes Basin," Ying Shao, Ross Lunetta, and team studied water quality in the Laurentian Great Lakes Basin. Corn acreage has expanded in response to high demand for corn as an ethanol feedstock (Volume I, Chapter 10). Here integrated remote sensing-derived products and the Soil and Water Assessment Tool (SWAT) within a GIS modeling environment were used to assess the impacts of cropland change on sediment yields within four selected watersheds using the system developed by US Environmental Protection Agency (EPA) scientists.

The SWAT models were calibrated over a six year period and predicted stream flows were validated. The R^2 values were 0.76, 0.80, 0.72, and 0.81 for the St. Joseph River, St. Mary's, the Peshtigo River, and the Cattaraugus Creek Watersheds, respectively. The corresponding E or Nash and Sutcliffe model efficiency coefficient values ranged from 0.24 to 0.79. The average annual sediment yields (tons/ha/yr) ranged from 0.12 to 4.44 for the baseline (2000–2008) condition. Sediment yields were predicted to increase for possible future cropland change scenarios. In freshwater systems, phosphorous in various forms is delivered downstream and is the limiting nutrient for water fertility and promotes algal growth diminishing water quality.

The first scenario was to convert all "other" agricultural row crop types to corn fields and switch the current/baseline crop rotation into continuous corn. The average annual sediment yields increased 7%–42% for different watersheds. The second scenario was to further expand corn plantings to hay/pasture fields. The average annual sediment yields increased 33%–127% compared to the baseline conditions.

This work demonstrated the utility of a multiple image data sets much like today's GEE and the prediction scenarios which can help in planning water quality interventions.

Volume I, Chapter 11, Climate and Precipitation Time Periods

In "Temporal downscaling of daily to minute interval precipitation by emulator modeling based genetic optimization," Venkatesh Budamala, Abhinav

Wadhwa, Amit B. Mahindrakar, and B. Srimuruganandam conducted a climate-related effort in the upper river Satluj basin of district Kinnaur, Himachal Pradesh (Volume I, Chapter 11). Data obtained from sensors or gauging stations provides many anomalies with respect to time and hence, it becomes complicated to produce the desired future climate data. The best alternative to predict and optimize the finer time period was to use Statistical Temporal Downscaling techniques. Based on the observed time period, the Statistical Temporal Downscaling identifies historical trends and provides feasible finer time scale data by integrating the incongruities in coarser time scale datasets.

In recent years, regression-based downscaling methods have been favored due to the identification of optimal climate occurrences with minimal complexity and computational burden. This sensitivity analysis helps to optimize model forecasting functionality.

3

Large-Scale Mapping and Monitoring Inland Waters by Google Earth Engine and Remote Sensing Techniques

Gordana Kaplan and Mateo Gašparović

CONTENTS

Introduction

The essence of life, water, has been uncontrollably used worldwide, making us reconsider the way we use and treat our water sources. Sustainable use of water is essential for the well-being of all creatures on the Earth and the Earth itself. With climate change and anthropogenic activities, water sources are becoming limited, and urgent action is needed. The starting point of preserving our water sources is through mindful management and sustainable actions. To achieve this, water sources need to be frequently monitored, mapped, and analyzed. While conventional methods for monitoring water areas are costly and challenging, remote sensing offers an alternate method.

Water body identification and extraction using satellite data has been a research topic ever since the beginning of remote sensing. Even though researchers have used different methods, such as emitted thermal radiation and active microwave emission, the reflected solar radiation has been widely used for water delineation (McFeeters, 1996). The very first index (Normalized Difference Water Index (NDWI)) for water extraction uses green light wavelengths to maximize the typical reflectance of water features and

DOI: 10.1201/9781003175018-3

near-infrared wavelengths to minimize the low reflectance by water areas [(Green–NIR)/(Green+NIR)].

The NDWI can be and has been easily applied over satellite images. Although NDWI can successfully delineate water and soil, over the years researchers have modifying NDWI, or have been investigating other indices for better results. Replacing the NIR band with shortwave NIR (SWIR), Xu (2006) modified the NDWI into MNDWI (modified NDWI) and achieved better results. However, the indices did not show satisfying results in mountainous and urban areas likely due to shadowing and slope lighting effects (Kaplan and Avdan, 2017a, Wang et al., 2020). Several indices have been proposed (Danaher and Collett, 2006, Feyisa et al., 2014, Fisher et al., 2016). The latest and most successful index is the Multi-Band Water Index (MBWI) (Wang et al., 2018b).

Simple band calculations are generally the most accurate way to extract water bodies in most cases. However, scientists have had challenges in extracting water bodies in dense urban (Wang et al., 2020) or mountainous areas (Kaplan and Avdan, 2017b). Thus, various methods have been developed for the same purpose. Besides water mapping, remote sensing data and techniques have been used for other water-related studies such as dynamics of active water bodies (Pickens et al., 2020, Mobariz and Kaplan, 2021), water quality (Avdan et al., 2019), mapping, and monitoring natural disasters caused by water bodies (Soltanian et al., 2019).

Mapping and monitoring water areas on a global or national scale requires large datasets, so most studies are conducted on small local areas (Nguyen et al., 2019). However, with the latest developments in the remote sensing community, limitations caused by processing of big data sets can be easily overcome using the Google Earth Engine (GEE) platform.

The success of GEE can be easily noticed through examination of the large number of studies conducted on the platform (Amani et al., 2020). The GEE platform uses a multi-petabyte catalog of satellite imagery and geospatial datasets generally used for Earth Observation purposes (Figure 3.1). This kind of data collection is suitable for multi-temporal monitoring of certain areas (Wang et al., 2018, Xia et al., 2019, Wang et al., 2020) and analyses of large scale, allowing scientists to obtain valuable information about the Earth.

GEE has also been used for water applications such as dynamics monitoring (Mobariz and Kaplan, 2020), surface water extraction (Jiang et al., 2021), flood mapping (Singh and Pandey, 2021), and long-term water changes analyses (Nguyen et al., 2019). A recent study (Avdan et al., 2019) used GEE to monitor the surface water dynamics over three decades in Turkey using Landsat imagery. Like this study, we will again emphasize the importance and advantages of GEE for Earth Observation purposes, with an accent on water bodies.

For this, we labored on mapping water bodies on a national scale over Croatia using one of European Space Agency's (ESA) finest spatial

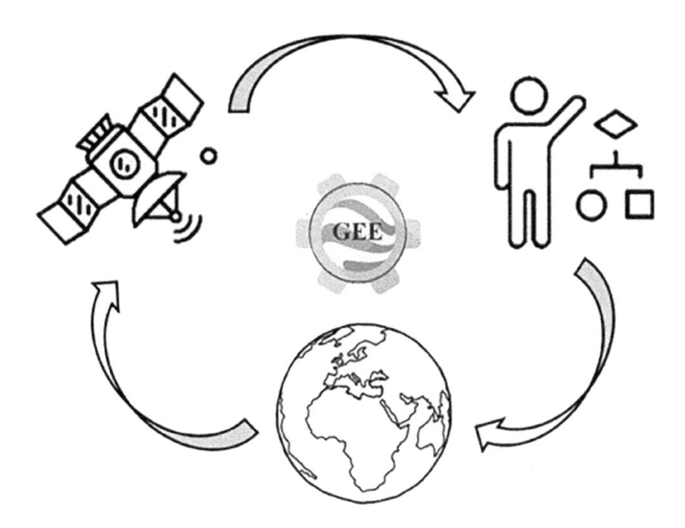

FIGURE 3.1
The GEE platform and processes for addressing geospatial datasets.

resolution satellites, Sentinel-2. With 10 m spatial and 6 days temporal resolution, Sentinel-2 has been used in different applications such as land cover mapping (Gašparović and Jogun, 2018), vegetation monitoring (Frampton et al., 2013, Addabbo et al., 2016), land transformations (Rashid et al., 2021), and wetland classification (Kaplan and Avdan, 2019). Also, Sentinel-2 data has been successfully used for water extraction (Kaplan and Avdan, 2017a).

Materials and Methods

The study area for national monitoring was the Republic of Croatia. Croatia is located at the crossroads of Central and Southeast Europe on the Adriatic Sea and has a population of 4.07 million. The capital of Croatia is Zagreb, and the entire country's territory covers 56,594 km^2 of land and 128 km^2 of water area. Most of Croatia has a moderately warm and rainy continental climate. The warmest areas of Croatia are at the Adriatic coast and especially in its immediate hinterland characterized by the Mediterranean climate, as the sea moderates the temperature highs.

Croatia belongs to the group of relatively water-rich countries. Problems with and around water have not yet worsened, and water resources are not yet a limiting factor for development. According to a 2003 UNESCO survey, Croatia ranks 5th in Europe and 42nd globally in water availability and richness.

Surface and groundwater balances show that Croatia has large, spatially unequally, and temporally distributed surface and groundwater. Accordingly, the institutions in charge of water management have the authority, obligation, and ability to design quality and harmonized solutions, sustainable for all parts of the water system and all water and water-dependent economy activities (Biondić et al., 2013).

In Croatia, there are 444 protected areas encompassing approximately 9% of the country. These include eight national parks (two of them are example subsets used in this research) and 12 nature parks. The most famous protected area and the oldest national park in Croatia is the Plitvice Lakes National Park, a UNESCO World Heritage Site (Radovic and Civic, 2006).

As previously mentioned, the entire Republic of Croatia was the study site for this research. Four example subsets were chosen for results best representing the inland waters mapping and monitoring at the national scale (Figure 3.2). Lake Vrana is a freshwater lake located in the center of Cres Island (1.5 km wide and 4.8 km long) and represents the first example subset (ES1, Figure 3.2b) with 5 km × 5 km dimensions. The second example subset (ES2, Figure 3.2c) with dimensions 15 km × 15 km is National Park Krka located along the middle-lower course of the Krka River in central Dalmatia, downstream from the Miljevci area and just a few km northeast of the city of Šibenik. The third example subset (ES3, Figure 3.2d) with dimensions of 10 km × 10 km spans the confluence of the Kupa and Sava Rivers and the City of Sisak is located 60 km southeast of Zagreb in the Sava basin. National Park Plitvice Lakes, founded in 1949 and located in the mountainous karst area of central Croatia, represents the fourth example subset (ES4, Figure 3.2e) with dimensions 3.5 km × 3.5 km.

Materials

The Copernicus Sentinel-2 mission, developed by the ESA, is a constellation of two twin satellites, Sentinel-2A and Sentinel-2B, launched in 2015 and 2017, respectively. The satellites are phased at 180° to each other, making the temporal resolution 6 days. Sentinel-2 carries a multispectral instrument payload collecting data in 13 different spectral bands with medium spatial resolution (Table 3.1) (Drusch et al., 2012). According to ESA, the main objectives of the Sentinel-2 mission are to provide high-resolution data with high revisit frequency, continuity with other multispectral sensors like SPOT and Landsat, and to provide data for Earth Observation applications. The temporal resolution of Sentinel-2 is nominally 6 days, and in order to cover the full area of Croatia 15 Sentinel-2 images are needed (Figure 3.3). As it is challenging to obtain cloud-free imagery from the full area, over a hundred Sentinel-2 images would have been required for the purpose of the study. This challenge has been overcome with GEE.

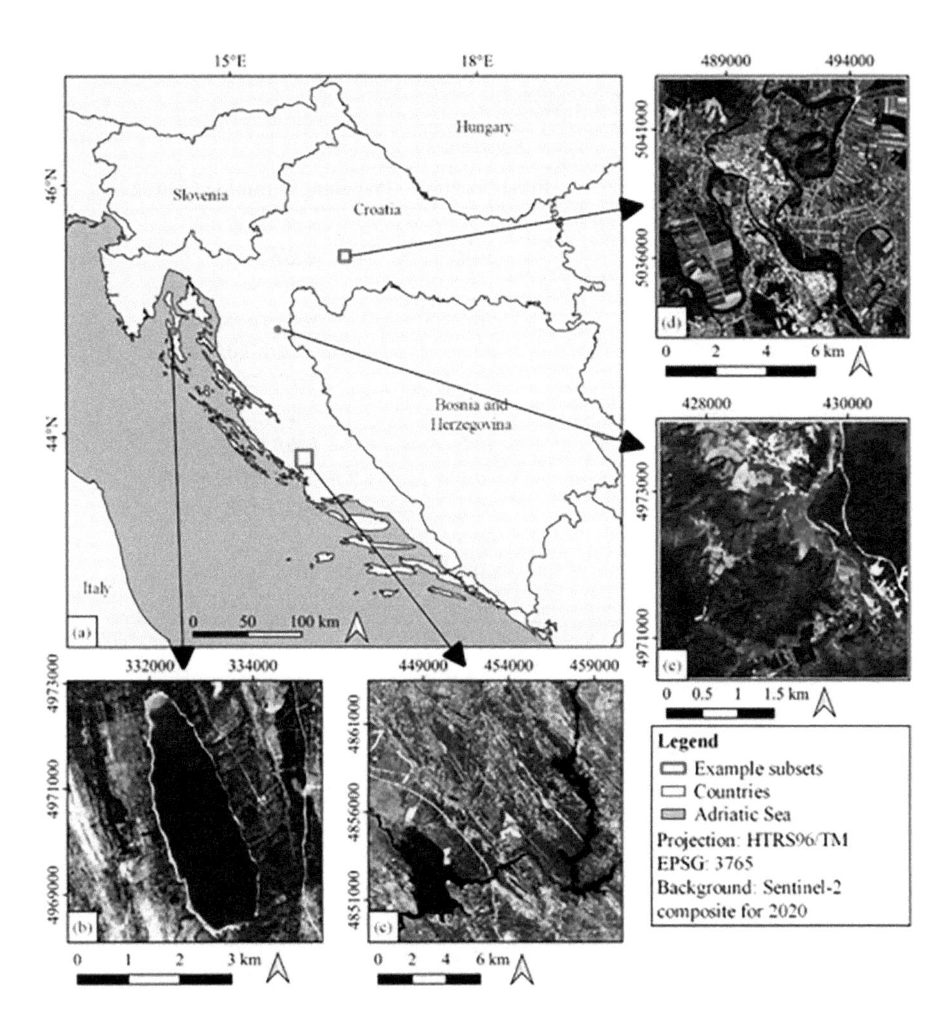

FIGURE 3.2
Inland waters mapping and monitoring at the national scale. (a) Regional graphic showing example subset positions of study areas. Images are from Sentinel-2. (b) Lake Vrana of Cres Island. (c) National Park Krka area located along the Krka River in Dalmatia. (d) The área at is the confluence of the Kupa and Sava Rivers and the City of Sisak. (e) The National Park Plitvice Lakes área. (f) The legend.

Methodology

To map and monitor water bodies on a national scale, according to the country's area, it can take dozens of Sentinel-2 satellite images to cover the study area. For instance, the area of Croatia can be covered with 15 Sentinel-2 images collected on different dates. Having satellite images collected on different dates would require significant image processing to lower the temporal difference between the images. However, in GEE, calculating medium value or

TABLE 3.1
Sentinel-2 Detailed Characteristics

	Sentinel-2	
Bands	**Pixel Size (m)**	**Wavelength (nm) (S2A)/(S2B)**
B1–Coastal	60	443.9/442.3
B2–Blue	10	496.6 /492.1
B3–Green	10	560/559
B4–Red	10	664.5/665
B5–Vegetation Red Edge	20	703.9/703.8
B6–Vegetation Red Edge	20	740.2/739.1
B7–Vegetation Red Edge	20	782.5/779.7
B8–NIR	10	835.1/833
B8A – Narrow NIR	20	864.8/864
B9—Water Vapor	60	945/943.2
B10–SWIR-Cirrus	60	1375
B11–SWIR	20	1613.7/1610.4
B12–SWIR	20	2202.4/2185.7

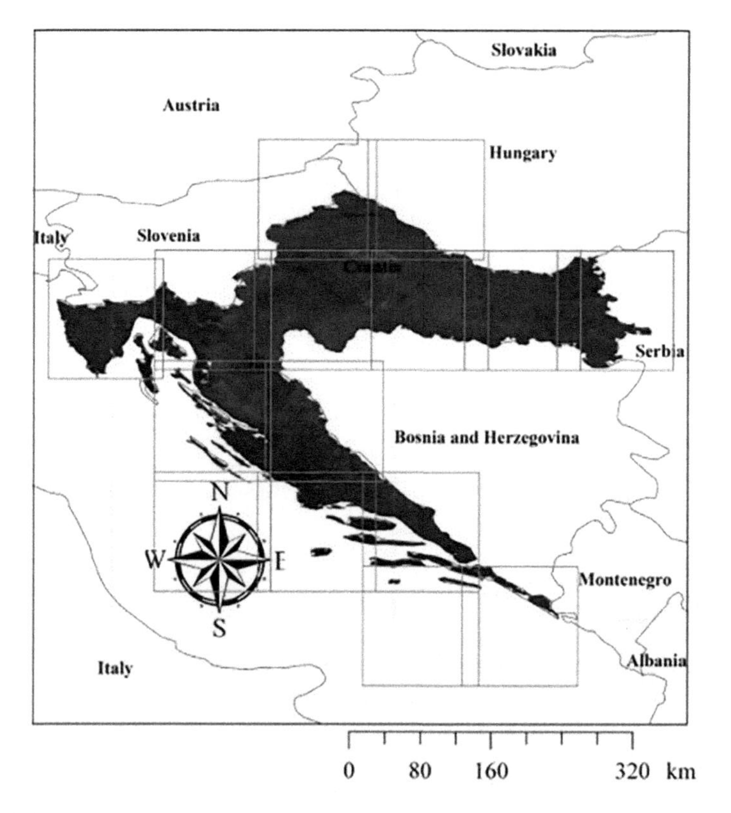

FIGURE 3.3
Sample of full area of Croatia covered in 15 Sentinel-2 images.

the total number of images to be collected is possible. This allows the user to obtain full coverage of satellite images over a specific area. The numerous possibilities in GEE, such as removing cloud cover from the images, make the platform full of advantages for Earth Observation.

The methodology developed for water extraction contains several steps. However, one of the main advantages of GEE is that after developing a certain methodology, with small modifications, it can be used in different study areas, with different data, and on different dates. Also, the results can be exported and shared with the public. The methodology of this study was performed entirely in GEE.

The methodology for water extraction is shown in Figure 3.4. Using Sentinel-2 satellite data, the dataset was first defined in the platform. All 10 and 20 m bands were loaded for better visualization of the imagery. Next, the Sentinel-2 data collection was pre-processed according to the area, date, and cloud mask filtering. Thus, the imagery was limited to Croatia's boundaries and the dates in the summer period in 2020, and a cloud filter mask of 15% was applied. Using this approach, a clear Sentinel-2 image collection over the area of Croatia was obtained (Figure 3.3).

For the water classification, MBWI was used as it showed the best results from the literature of the index-based methods. MBWI is based on the differences between water and other low reflectance surfaces, limiting the ranges of brightness values employed to those at the lower or "darker" portion of the terrestrial spectral range characteristic of water. The MBWI is designed to restrain non-water pixels and enhance surface water information. Detailed descriptions of the formulation can be found in Wang et al. (2018a). While normal NDWI is constructed from only two bands, MBWI uses five bands, green, red, NIR, SWIR1, and SWIR2.

$$\text{MBWI} = 2 \times \text{GREEN} - \text{RED} - \text{NIR} - \text{SWIR1} - \text{SWIR2} \tag{3.1}$$

The index was applied to the Sentinel-2 images. After the calculation of MBWI, visual inspection was made to determine the suitable threshold of the index for the study area. Using the threshold, the pixels in the study area have been separated into two classes, water and other.

In remote sensing, accuracy assessment is crucial for reliability of the results. Even though assessment through statistical analysis is important, visual inspection has also been widely used for validation of the results. As the number of inland water bodies is not large, visual analysis has been performed to evaluate the water extraction results over Croatia. In addition, after the classification and the accuracy assessment, change detection experiment between several water bodies was performed between 2017 and 2019 which demonstrated good results.

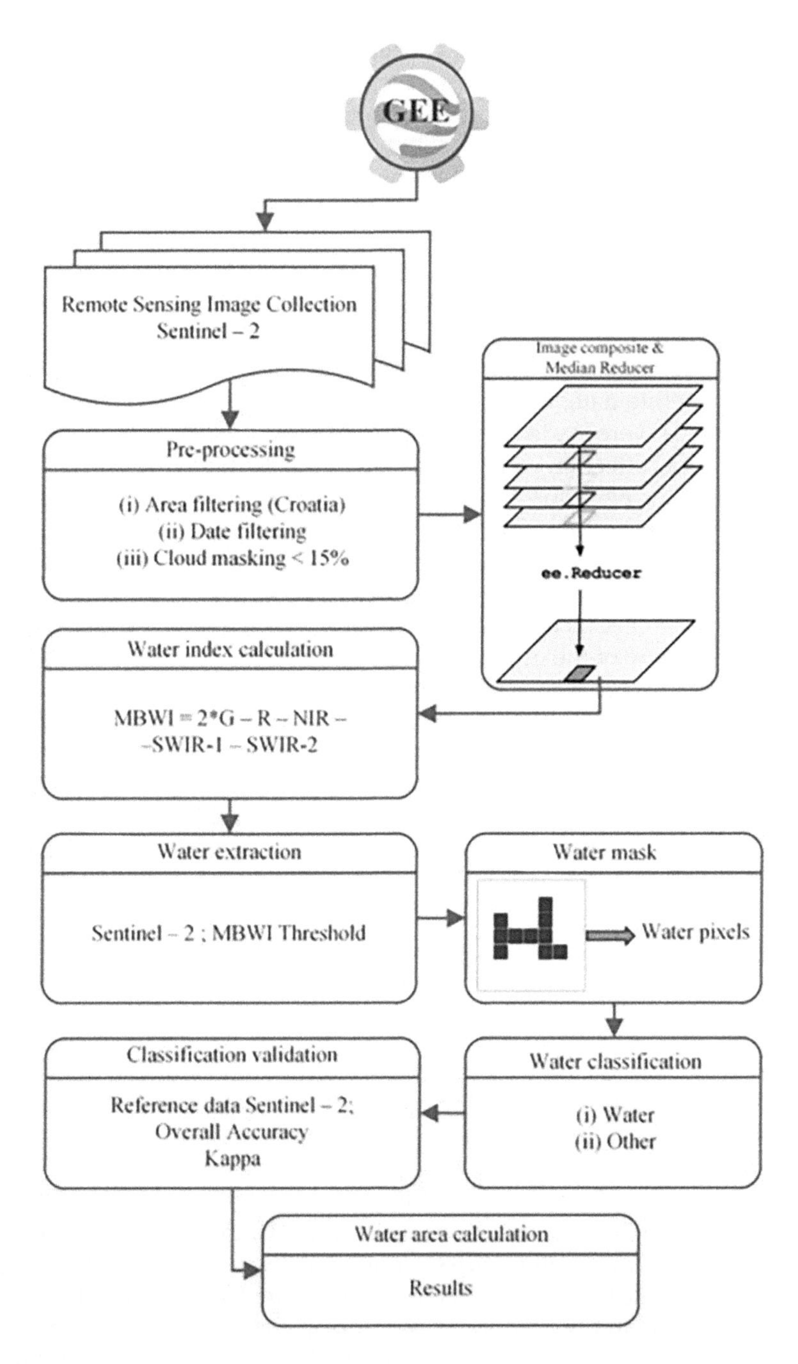

FIGURE 3.4
Presents a flow chart of processes used in the study.

Results

As a result, a national map of the inland water bodies over Croatia from 2020 has been produced. As the inland water bodies are relatively small, the results have been presented over smaller samples in Figures 3.5 and 3.6. Figure 3.5 represents four sub-areas from several locations in Croatia as a true color composite from Sentinel-2 data in the summer of 2020. We aimed at presenting different shapes of water bodies, namely rivers, dam reservoirs, smaller lakes, and lakes in national parks.

The most challenging areas for extracting water bodies were the urban areas, where shadows from high buildings can be mistakenly classified as water bodies (Kaplan and Avdan, 2017a). However, the methodology used in this study successfully managed to extract water bodies in urban areas (Figures 3.5 and 3.6, ES-3). For the same sub-areas in Figure 3.5, we tried to do change detection for the years 2017 and 2020; however, as seen in Figure 3.6, no significant change has been noticed in these areas.

The results of the water mapping over Croatia for summer 2020 are accessible in GEE through the links (https://kaplangorde.users.earthengine.app/view/water, last checked February 16, 2022).

Spatiotemporal changes in water bodies have been primarily investigated using remote sensing data (Sarp and Ozcelik, 2017, Topouzelis et al., 2017). Besides yearly changes of water bodies, researchers have also used remote sensing data for hazard evaluation in flooded areas (Sharma et al., 2019, Huang and Jin, 2020). Below we give an example of mapping and monitoring flooded areas using the developed algorithm in GEE with Sentinel-2 data. The sub-area was flooded in January 2020, and the event has been investigated in another scientific research effort (Gašparović and Klobučar, 2021). The visual results can be clearly seen in Figure 3.6, where the flooded areas have been mapped.

Discussion and Conclusions

In the summer of 2021, natural disasters, such as wildfires and floods, have occurred all over the World. Rapid and accurate mapping and monitoring of pre- and post-disaster conditions are gaining more and more attention. Currently, remote sensing is the optimal source of data for rapid mapping and updating land cover changes. Also, GEE is gaining more attention from researchers in different fields. GEE can provide open-source data and is also a user-friendly application. With a 10-m spatial resolution and 6-day temporal (revisit) resolution, Sentinel-2 offers middle scale resolution and it is optimal for environmental studies. Here, we have emphasized the basics of

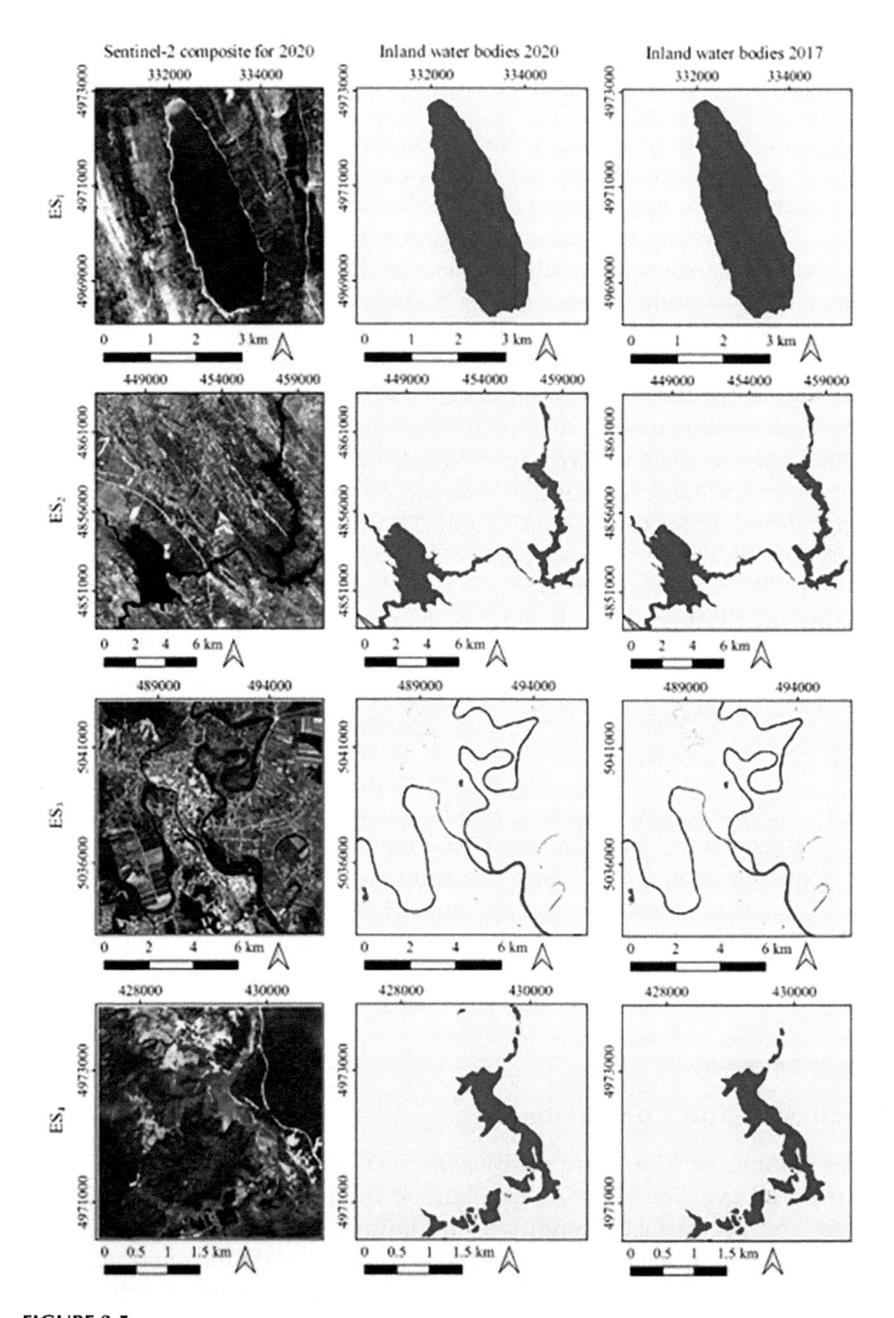

FIGURE 3.5
Four sub-areas from several locations in Croatia as a true color composite from Sentinel-2 data. (a) Lake Vrana composite. (b) Lake Vrana area in 2020. (c) Lake Vrana area in 2017. (d) National Park Krka area composite. (e) Krka area in 2020. (f) Krka area in 2017. (g) Kupa and Sava Rivers and Sisak composite. (h) Sisak area in 2020. (i) Sisak area in 2017. (j) The National Park Plitvice Lakes área composite. (k) Plitvice Lakes area in 2020. (l) Plitvice Lakes area in 2017.

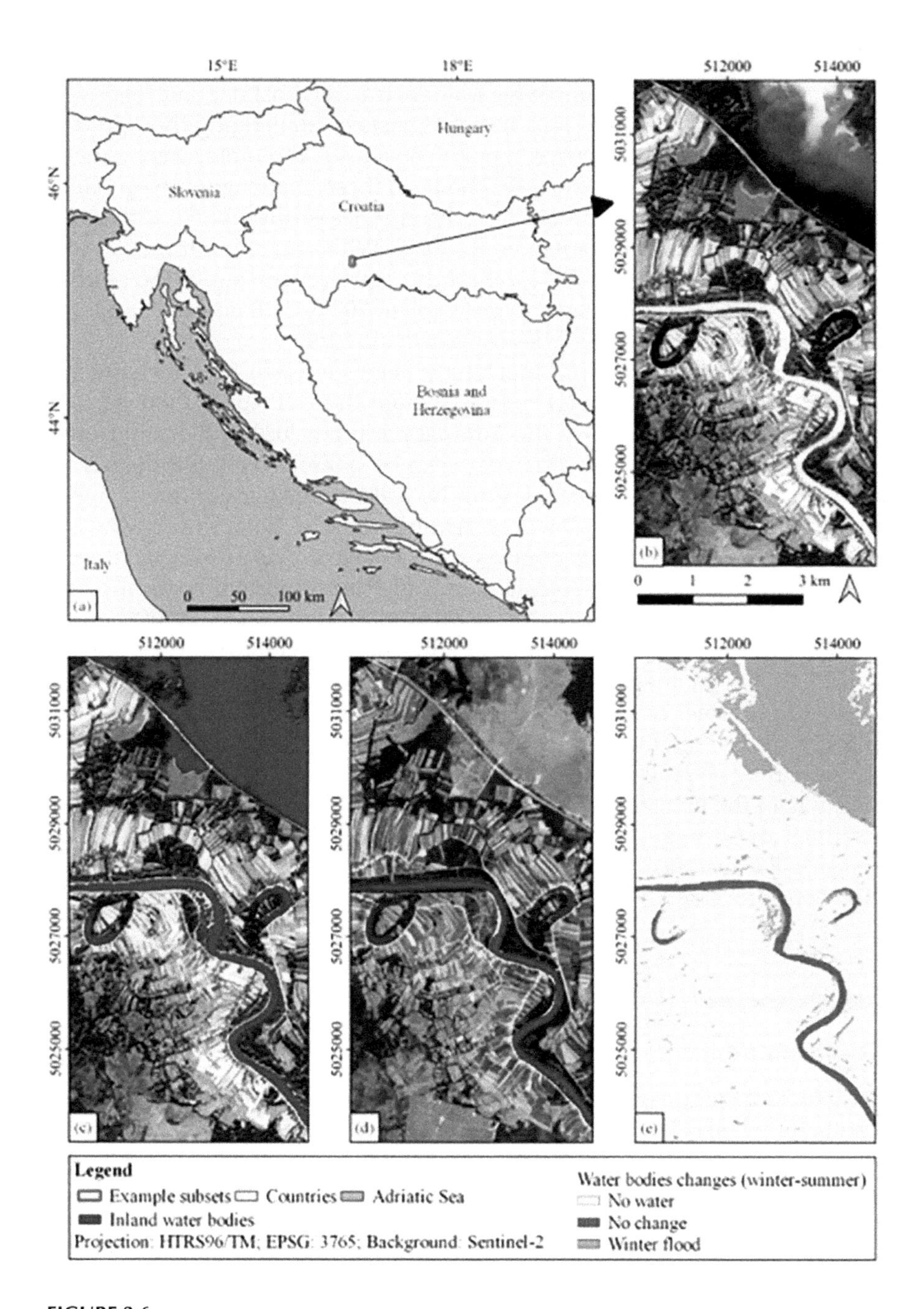

FIGURE 3.6
The flooded areas have been mapped. (a) Location of study area. (b) Composite image. (c) Water areas in blue or medium grey. (d) Flood areas in blue or medium grey on composite. (e) Flood areas highlighted on background. (f) The legend.

remote sensing, the importance of GEE for rapid and accurate mapping and monitoring through an example for water extraction on a national basis.

Mapping water resources on a national basis is challenging through conventional on-site methods. Even though conventional remote sensing methods did lower the cost and time for the task, these challenges are overcome with GEE. Remote sensing data have been successfully linked with gauging station data as well (Kaplan et al., 2020). While mapping three decades of water change on a national level requires pre-processing of the satellite imagery and analysis for every year (Kaplan, 2020), GEE enables rapid and accurate mapping of water bodies (Avdan et al., 2019).

The inland water bodies in Croatia have been successfully extracted using the algorithm developed in GEE. Even though the GEE application is based on the Sentinel-2 images from the summer period in 2020, with small adjustments, analysis for the previous years can be done as well. For the period between 1984 and 2011, Landsat-5 can be used instead of Sentinel-2, while after 2013, the analyses can be done using Landsat-8 data. In this way, the researchers can analyze water bodies since 1984 using open-source data.

As mentioned here, change detection is also an important application for wide-ranging and/or near-time temporal applications. Following the natural disasters, many researchers and local administrations have applied remote sensing for post-damage assessment and updated the situation of the affected areas (e.g., summer of 2021). The developed application in GEE can be useful in mapping and monitoring flooded areas and drought monitoring.

It should be noted that GEE is an excellent source for handling big satellite data, where the imagery can be processed in the cloud, saving the researchers time and energy from downloading and pre-processing the satellite imagery. For instance, in this study, the number of Sentinel-2 data has been easily pre-processed in the GEE platform.

Recommendations

Even though middle-resolution satellite imagery gives timely and accurate results, the need for higher temporal and spatial resolution, especially in smaller areas, is inevitable. Recently developed satellite sensors, such as PlanetScope, with their average daily revisit in most parts of the World, are promising data for solving problems. Besides, scientists have integrated Sentinel-2 with PlanetScope data to achieve higher spatial, temporal, and spectral resolution. For future studies, we recommend integrating different sensors and sensor platforms to investigate the possibilities that the fused data may open.

The results of the water mapping over Croatia for summer 2020 are accessible in GEE through the following link: https://kaplangorde.users.earthengine.app/view/water (last checked February 16, 2022).

References

Addabbo, P., Focareta, M., Marcuccio, S., Votto, C. and Ullo, S. L. 2016. Contribution of Sentinel-2 data for applications in vegetation monitoring. *ACTA IMEKO*, 5(2), 44.

Amani, M., Ghorbanian, A., Ahmadi, S. A., Kakooei, M., Moghimi, A., Mirmazloumi, S. M., Moghaddam, S. H. A., Mahdavi, S., Ghahremanloo, M. and Parsian, S. 2020. Google Earth Engine cloud computing platform for remote sensing big data applications: A comprehensive review. *IEEE Journal of Selected Topics in Applied Earth Observations and Remote Sensing*, 13, 5326–5350.

Avdan, Z., Kaplan, G., Goncu, S. and Avdan, U. 2019. Monitoring the water quality of small water bodies using high-resolution remote sensing data. *ISPRS International Journal of Geo-Information*, 8, 553.

Biondić, D., Holjević, D. and Petraš, J. 2013. Floods in the Danube River Basin in Croatia in 2010. In Loczy, D. (ed.) *Geomorphological Impacts of Extreme Weather*. Springer Geography, Dordrecht, pp. 141–153.

Danaher, T. and Collett, L. 2006. Development, optimisation and multi-temporal application of a simple Landsat based water index. In *Proceeding of the 13th Australasian Remote Sensing and Photogrammetry Conference*, Canberra, Australia.

Drusch, M., Del Bello, U., Carlier, S., Colin, O., Fernandez, V., Gascon, F., Hoersch, B., Isola, C., Laberinti, P. and Martimort, P. 2012. Sentinel-2: ESA's optical high-resolution mission for GMES operational services. *Remote Sensing of Environment*, 120, 25–36.

Feyisa, G. L., Meilby, H., Fensholt, R. and Proud, S. R. 2014. Automated water extraction index: A new technique for surface water mapping using Landsat imagery. *Remote Sensing of Environment*, 140, 23–35.

Fisher, A., Flood, N. and Danaher, T. 2016. Comparing Landsat water index methods for automated water classification in eastern Australia. *Remote Sensing of Environment*, 175, 167–182.

Frampton, W. J., Dash, J., Watmough, G. and Milton, E. J. 2013. Evaluating the capabilities of Sentinel-2 for quantitative estimation of biophysical variables in vegetation. *ISPRS Journal of Photogrammetry and Remote Sensing*, 82, 83–92.

Gašparović, M. and Jogun, T. 2018. The effect of fusing Sentinel-2 bands on land-cover classification. *International Journal of Remote Sensing*, 39, 822–841.

Gašparović, M. and Klobučar, D. 2021. Mapping floods in lowland forest using Sentinel-1 and Sentinel-2 data and an object-based approach. *Forests*, 12, 553.

Huang, M. and Jin, S. 2020. Rapid flood mapping and evaluation with a supervised classifier and change detection in Shouguang using sentinel-1 SAR and sentinel-2 optical data. *Remote Sensing*, 12, 2073.

Jiang, Z., Jiang, W., Ling, Z., Wang, X., Peng, K. and Wang, C. 2021. Surface water extraction and dynamic analysis of Baiyangdian Lake based on the Google Earth Engine platform using Sentinel-1 for reporting SDG 6.6. 1 indicators. *Water*, 13, 138.

Kaplan, G. 2020. Mapping three decades water changes in North Macedonia using remote sensing data. *Micro, Macro and Mezzo Geoinformation*, 15, 45–53.

Kaplan, G. and Avdan, U. 2017a. Object-based water body extraction model using Sentinel-2 satellite imagery. *European Journal of Remote Sensing*, 50, 137–143.

Kaplan, G. and Avdan, U. 2017b. Water extraction technique in mountainous areas from satellite images. *Journal of Applied Remote Sensing*, 11, 046002.

Kaplan, G. and Adan, U. 2019. Evaluating the utilization of the red edge and radar bands from sentinel sensors for wetland classification. *Catena*, 178, 109–119.

Kaplan, G., Avdan, Z., Avdan, U. and Jovanovska, T. 2020. Monitoring shared international waters with remote sensing data. *Resilience*, 4, 77–88.

Mcfeeters, S. K. 1996. The use of the Normalized Difference Water Index (NDWI) in the delineation of open water features. *International Journal of Remote Sensing*, 17, 1425–1432.

Mobariz, M. A. and Kaplan, G. 2020. Monitoring Amu Darya River channel dynamics using remote sensing data in Google Earth Engine. *Geodetski List*, DOI: 10.3390/ECWS-5-08012.

Mobariz, M. A. and Kaplan, G. 2021. Moving borders: Mapping and monitoring Amu Darya River dynamics using remote sensing data and techniques. *Geodetski List*, 1, 29–44.

Nguyen, U. N., Pham, L. T. and Dang, T. D. 2019. An automatic water detection approach using Landsat 8 OLI and Google Earth Engine cloud computing to map lakes and reservoirs in New Zealand. *Environmental Monitoring and Assessment*, 191, 1–12.

Pickens, A. H., Hansen, M. C., Hancher, M., Stehman, S. V., Tyukavina, A., Potapov, P., Marroquin, B. and Sherani, Z. 2020. Mapping and sampling to characterize global inland water dynamics from 1999 to 2018 with full Landsat time-series. *Remote Sensing of Environment*, 243, 111792.

Radovic, J. and Civic, K. 2006. *Biodiversity of Croatia*. State Institute for Nature Protection, Ministry of Culture, Republic of Croatia.

Rashid, M. B., Habib, M. A., Khan, R. and Islam, T. 2021. Land transform and its consequences due to the route change of the Brahmaputra River in Bangladesh. *International Journal of River Basin Management*, 1–13.

Sarp, G. and Ozcelik, M. 2017. Water body extraction and change detection using time series: A case study of Lake Burdur, Turkey. *Journal of Taibah University for Science*, 11, 381–391.

Sharma, T. P. P., Zhang, J., Koju, U. A., Zhang, S., Bai, Y. and Suwal, M. K. 2019. Review of flood disaster studies in Nepal: A remote sensing perspective. *International Journal of Disaster Risk Reduction*, 34, 18–27.

Singh, G. and Pandey, A. 2021. Mapping Punjab flood using multi-temporal open-access synthetic aperture radar data in Google Earth Engine. In: Pandey, A., Mishra, S., Kansal, M., Singh, R., Singh, V.P. (eds) *Hydrological Extremes*. Water Science and Technology Library, vol 97, p. 75–85. Springer.

Soltanian, F. K., Abbasi, M. and Bakhtyari, H. R. 2019. Flood monitoring using NDWI and MNDWI spectral indices: A case study of Aghqala flood-2019, Golestan Province, Iran. *The International Archives of Photogrammetry, Remote Sensing and Spatial Information Sciences*, 42, 605–607.

Topouzelis, K., Papakonstantinou, A. and Doukari, M. 2017. Coastline change detection using unmanned aerial vehicles and image processing technique. *Fresenius Environmental Bulletin*, 26, 5564–5571.

Wang, C., Jia, M., Chen, N. and Wang, W. 2018a. Long-term surface water dynamics analysis based on Landsat imagery and the Google Earth Engine platform: A case study in the middle Yangtze River Basin. *Remote Sensing*, 10, 1635.

Wang, X., Xie, S., Zhang, X., Chen, C., Guo, H., Du, J. and Duan, Z. 2018b. A robust Multi-Band Water Index (MBWI) for automated extraction of surface water from Landsat 8 OLI imagery. *International Journal of Applied Earth Observation and Geoinformation*, 68, 73–91.

Wang, Y., Li, Z., Zeng, C., Xia, G.-S. and Shen, H. 2020. An urban water extraction method combining deep learning and Google Earth engine. *IEEE Journal of Selected Topics in Applied Earth Observations and Remote Sensing*, 13, 768–781.

Xia, H., Zhao, J., Qin, Y., Yang, J., Cui, Y., Song, H., Ma, L., Jin, N. and Meng, Q. 2019. Changes in water surface area during 1989–2017 in the Huai River Basin using Landsat data and Google earth engine. *Remote Sensing*, 11, 1824.

Xu, H. 2006. Modification of normalised difference water index (NDWI) to enhance open water features in remotely sensed imagery. *International Journal of Remote Sensing*, 27, 3025–3033.

4

Identification of Potential Runoff Storage Zones within Watersheds

Vikas Kumar Rana and Tallavajhala Maruthi Venkata Suryanarayana

CONTENTS

DOI: 10.1201/9781003175018-4

Introduction

A naturally occurring geohydrological unit draining to a common point by a system of natural streams/drains is defined as a watershed. Integrated watershed management is a multidisciplinary approach for rational utilization of natural resources existing in the watershed. The water resources of many countries are faced with the growing pressure of population and rapidly growing urban areas that result in water deficiency and food insecurity.

Water resources are used by human societies for both consumptive and non-consumptive purposes in households, agriculture, municipalities and industries. Human activities, on the other hand, have a significant impact on global water supply demand (UNESCO-WWAP, 2009). Further, despite supply constraints caused by population growth, the pursuit of higher standards of living and economic growth, demand for water continues to rise. Agriculture, industry and municipalities account for roughly 70%, 20% and 10% of global water use, respectively (IWMI, 2007).

To feed the world's growing population, more land must be irrigated, implying that more water is required. The amount of irrigated land is expected to increase by more than 40% by 2080 (Fischer et al., 2007). Domestic and municipal water demand is also increasing. By 2014, the average global availability of renewable freshwater resources had fallen to less than 6,000 m^3 per person per year, a reduction of almost 40% since the 1970s. Furthermore, freshwater resources are unevenly distributed around the world and are affected by strong seasonality; as global demand for water continues to rise by about 1% per year, available resources are depleted further (de Castro-Pardo et al., 2021).

Water scarcity has become a serious problem in several parts of the world, especially in developing nations like India (Kumar and Jhariya, 2017). In India, ever-increasing population exerts enormous pressure on water resources of which per capita water availability is decreasing day by day (Singh et al., 2017). Hence, it becomes necessary to maximize possible water sources within the watershed. There is a growing need for cost-effective and time-saving methods to identify areas that are suitable for water storage. Before execution, water storage structures require significant investment, and hence, it is important to identify the potential runoff storage zones for these structures.

The need for alternative and reliable sources of potential runoff and water storage zones in a catchment are the zones where rainwater-harvesting structures can be successful; as like small dams or check dams, nala bunds, gully plug, bundhis (local name in India), percolation tanks, etc. can be constructed in a planned and systematic manner. Thereby creating water buffers within the catchment, which will help reduce vulnerability to drought and seasonal variations in rainfall. These structures can be used for multiple purposes, such as agriculture, livestock watering and domestic use. In general, check dams are constructed on lower-order streams, and the slope of the terrain should be from flat to gentle to retain maximum quantity of water with less height of a check dam. It should be located nearer to agricultural areas and settlements to convey the water. Check dams have greater importance than other structures since they have a complimentary benefit of controlling soil erosion.

Nala bunds and percolation tanks are structures constructed across or nearer to nalas (streams) for checking velocity of runoff, increasing water percolation, increasing soil-moisture regime and to hold the silt flow which would otherwise reach the multipurpose reservoirs and reduce their useful life. Nala bunds are less expensive, smaller in dimension and constructed using locally available material whereas percolation tanks are larger and more expensive than nala bunds.

The feasibility of sites for locating percolation tanks depends upon technical and economic considerations such as the sites should be selected in a relatively flat nala reach and the slope of the nala should preferably not be more than 2%. Water harvesting bundhis are almost like minor irrigation tanks except that they do not have extensive canal systems and their command area is limited to the fields downstream. They are used to collect and impound surface runoff during monsoon rains. This facilitates infiltration to raise groundwater levels in the zone of influence of the bundhi and provides irrigation in the fields lying in proximity. Water harvesting bundhi also moderates the peak flows, partly by storing and partly through flooding.

Role of Remote Sensing and Geographic Information System

Remote sensing and geographic information systems (GIS) together fulfil the requirement of identification of potential runoff storage zones in a catchment by providing a conceptual framework for collecting and analysing spatial and non-spatial data (Krois and Schulte, 2014). The use of geospatial data in conjunction with GIS techniques has made it simple to create databases pertaining to an area's hydrological potential. Proper selection of factors is of great importance for identifying sites for specific water storage structures. A review of literature by Ammar et al. (2016) found that a different number of data layers have been used by researchers to capitalize on the availability of data for potential areas or to identify sites suitable for rainwater harvesting. In several studies related to the identification of water storage sites, the weighted linear combination technique has been used for the integration of biophysical layers in a GIS environment. Weerasinghe et al. (2011) focused on using a GIS and remote sensing (RS) and developed a spatial analysis model named Geographic Water and Management Potential. The model was able to find potential water harvesting and storage sites for water storage and soil-moisture conservation on farms. In most studies, a range of weights was decided arbitrarily or through personal experience, or weights were assigned on the scale of 1–5 or 1–100, whereas only a few studies assigned weights on a standard 1–9 scale, as suggested by Saaty (1987). De Winnaar et al. (2007) conducted a study in which the SCS-CN method was applied to identify potential runoff-harvesting sites in a small sub-catchment in South Africa. The inputs included socio-economic data gathered from available sources and from field surveys, a digital elevation model with 20 m resolution to extract slope information, a soil survey providing soil data, digital images and aerial photographs. Similarly, Ghani et al. (2013) explored potential rainwater storage sites by examining runoff patterns using a hydrologic model with the GIS/RS approach. A 90 m digital elevation model was used as a source for catchment elevation data to determine flow direction, drainage lines and runoff. Krois and Schulte (2014) presented a GIS and a multi-criteria evaluation approach to identify and rank sites for the implementation of soil and water conservation techniques within the Ronquillo watershed. Criteria maps were created by reclassifying the spatial maps based on the suitability level for each RWH technique. A pairwise comparison matrix method (analytic hierarchy process (AHP)) calculated the relative-importance weight of each criterion for each rainwater-harvesting technique. The weighted overlay process in the GIS determined the suitability maps for each rainwater-harvesting technique. Rainfall, runoff coefficient, slope, land use, soil texture and soil depth were selected based on the Food and Agricultural Organization (FAO) guidelines. The assessment of the dominance of one criterion over another was based on the authors' expertise and a literature survey.

Generally, the GIS is based on the use of decisive rules in a site selection approach to decide how a collection of parameter maps may be fused together to ensure that alternative sites are designed in accordance with several assessment criterion preferences. In GIS, the most used decision judgements are the weighted linear combination and the Boolean overlay procedures. The weighted linear combination is a method for combining maps that assigns a normalized score to each category of a specific criterion and a factor weight to the criteria itself. Because of its versatility in determining appropriate sites, this approach provides superior site selection. The factors for the Boolean overlay method might be true or false. The Boolean overlay technique selects sites for rainwater harvesting based on whether the AND or OR operations are used. When using the AND operation, the selection of rainwater harvesting (RWH) sites is limited to small, dispersed locations. This strategy is useful for omitting a particular area indicated by the weighted linear combination method.

The continued use of GIS and RS in site selection has drawn more attention to these technologies. Other relevant tools, such as the AHP, have been recommended for weighing thematic spatial information and creating runoff potentials in site-suitability studies, in addition to GIS, RS and SCS-CN approaches (Singh et al., 2009; Eshghizadeh et al., 2018).

AHP is a multi-criteria decision-making tool that provides a structured technique for organizing and analysing complex decisions based on mathematics and expert knowledge (Saaty, 2008). Several studies have reported value for site-suitability determinations using Multi Criteria Decision Making (MCDM) and AHP in GIS environments (Al-Adamat, 2008; Kahinda et al., 2008; Pauw et al., 2008; Mahmoud and Alazba, 2014). AHP is a popular weighting method in the field of MCDM (Saaty, 1977; Rozos et al., 2011; Karimi and Zeinivand, 2019).

The AHP is a theory of measurement through a pairwise comparison matrix and relies on the judgements of experts to derive priority scales. It is used as a higher cognitive process tool to determine the percentage importance of various criteria in the determination of suitable sites. It consists of three main phases: construction of the hierarchy, priority analysis of data, and confirmation of consistency. According to Ammar et al.'s (2016) review study on identification of suitable sites for water storage in arid and semi-arid regions, it was found out that the most common biophysical layers or criteria applied were slope followed by land use/land cover and soil type. We aim to improve the existing methodology by introducing new layers along with the commonly used biophysical layers for estimation of potential runoff storage zones to identify water storage sites.

Case Study

In the present study, a GIS-based conceptual framework was applied with the MCDM technique using AHP to produce suitability maps of potential runoff storage zones within the watershed. The conceptual framework will help to identify potential runoff storage zones for water storage sites based on the various physical characteristics (Rainfall, Slope, Land use/land cover, Height above the nearest drainage (HAND), Stream order, Curve number, Topographic wetness index (TWI)) of the Vishwamitri watershed. This will help concerned authorities in the proficient execution of water-related plans and schemes, improve water shortage conditions, reduce dependability on ground water and ensure sustainable water availability for local and agricultural purposes.

The present study was conducted in the Vishwamitri Watershed in Vadodara district, Gujarat. Vadodara district is located at the south of the Tropic of Cancer in the transition zone of heavy rainfall areas of South Gujarat and arid areas of the North Gujarat plains. It has a subtropical climate with moderate humidity and forms a part of the great Gujarat plain. The eastern portion of the district is hilly terrain with several ridges, plateaus and isolated relict hills with an elevation of 150–481 m above the mean sea level. The Vishwamitri river originates from the hills of Pavagadh, 43 km northeast of Vadodara. The Pavagadh hills are made of trappean rocks that emerge abruptly 830 m above the mean sea level. The Vishwamitri river has a channel length of around 70 km, of which 58 km flows through Vadodara district. It meets the Dhadhar river at Pingalwada in Vadodara district. Figure 4.1 shows the geographical location of the study area.

FIGURE 4.1
Geographical location of the study area.

Identification of Potential Runoff Storage Zones

To find the potential runoff storage zones, the workflow was divided into four steps (Figure 4.2). Firstly, the rainfall analysis was carried out using SPI and annual rainfall. Secondly was the processing of spatial data and creation of spatial data layers. Thirdly, criteria weights were determined using AHP. Lastly was the executing weighted overlay process (WOP) within the GIS.

Rainfall Analysis

The study area falls under the arid areas of the north Gujarat plains. Potential runoff storage zones or structures require considerable rainfall and hence it is important to analyse variability of rainfall within the watershed for the suitable locations of these storage zones or structures. For rainfall variability analysis, two indicator, viz., annual rainfall and Standardized Precipitation Index (SPI),

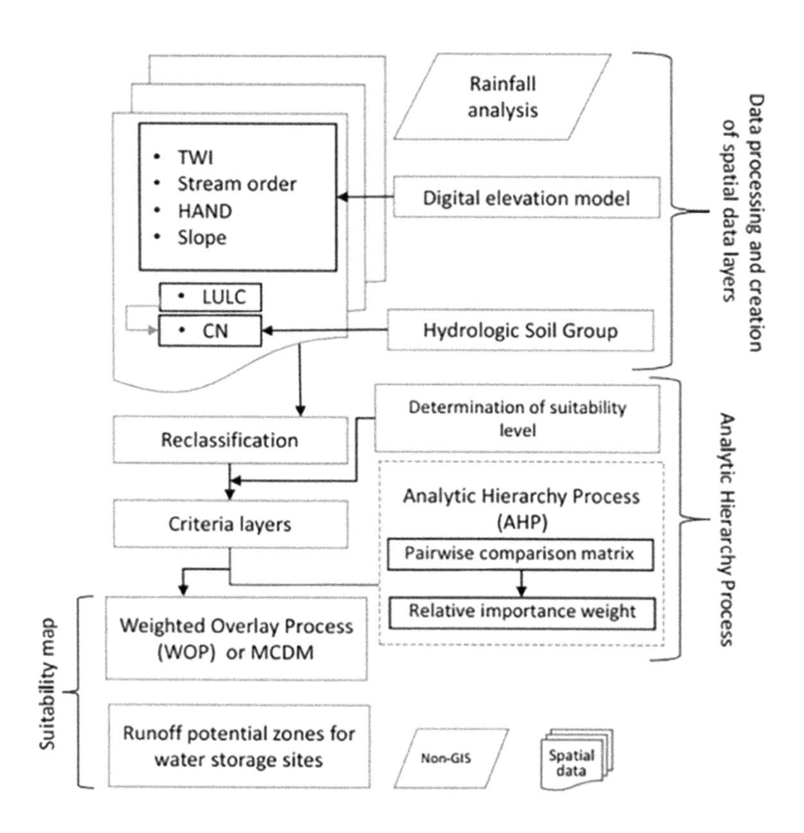

FIGURE 4.2
Multi criteria decision-making (MCDM) technique workflow using AHP for identification of potential runoff storage zones for water storage.

have been used. Annual rainfall is highly influenced by the amount and intensity of rainfall, frequency of occurrence and areal distribution as well as timing.

The approach examines the annual rainfall along with the SPI drought index application for the Vishwamitri watershed and it was calculated from historical precipitation data. Results of SPI and annual rainfall analyses help in evaluating whether the area is suitable for water storage structures or not. Favourable results qualified the area for identification of water storage.

The SPI calculation for any location is based on the long-term precipitation record for a desired period. Precipitation is normalized using a probability distribution function and allows for the estimation of both dry and wet periods. Daily rainfall data were collected by the State Water Data Centre, Gandhinagar, Gujarat, for a total of 56 years (1961–2016) and from four rain gauge stations Vadodara, Padra, Savli and Waghodia. These were used for computation of SPI and annual rainfall. The annual SPI classification system used by McKee et al. (1993) is shown in Table 4.1, and it was computed as

TABLE 4.1

Annual Standardized Precipitation Index

Classification	SPI
Near normal	–0.99 to 0.99
Moderately wet years	1.0–1.49
Moderately dry years	–1.0 to –1.49
Very wet	1.5–1.99
Severely dry years	–1.5 to –1.99
Wet extreme	≥+2.0
Dry extreme	≤–2.0

described by Akinsanola and Ogunjobi (2014) and Adegoke and Sojobi (2015). Positive SPI values indicate greater than median precipitation, and negative values indicate less than median precipitation. Drought starts when the SPI value is equal or below –1.0 and ends when the value becomes positive.

$$\mathrm{SPI} = \frac{X - \bar{X}}{\sigma}$$

X = rainfall in each particular year

$\bar{X}$ = mean rainfall in each particular year

σ = the standard deviation of rainfall in each particular year

Processing and Creation of Spatial Data Layers

Topographic Wetness Index (TWI)

The TWI was first introduced by Beven and Kirkby (1979). TWI is a widely used topographically based soil wetness model that identifies wet areas. It assumes that local topography controls the movement of water in slopped terrain which quantifies the effect of the local topography on runoff generation. The index is represented as the natural logarithm of the ratio of upslope flow accumulation area and slope at the cell.

$$\mathrm{TWI} = \mathrm{Ln}(\text{flow accumulation} + 1) / \left(\tan\left(\left((\text{slope in degrees})3.14\right)/180\right)\right)$$

Topographic wetness at a particular point on the landscape is the ratio between the catchment area contributing to that point and the slope at that point (Wilson and Gallant, 2000). Locations with a high TWI value typically have large upslope areas and are expected to have higher water availability. On the other hand, locations with small TWI values have small upslope areas that are assumed to have lower water availability. Also, Steep locations receive a small TWI value and are expected to be better drained than gently sloped locations, which receive a high TWI value (Sørensen and Seibert, 2007; Ågren et al., 2014; Bjelanovic, 2016; Hojati and Mokarram, 2016; Loritz et al., 2019).

The TWI calculation employed the Topography Toolbox of ArcGIS 10.1 (Dilts, 2015). First, the DEM was pre-processed to remove shallow sinks and the potential impact of model artefacts in further analysis. The second step included calculation of prerequisites for further TWI computations of slope and catchment area. The latter parameter was calculated using the Multiple Flow-Direction method (Quinn et al., 1991) and Generation of slope map using Topography Position Index (TPI).

The TPI is the basis of the landform classification system. Gallant and Wilson (2000) defined TPI as the relative topographic position of a central point as the difference between the elevation at this point and the mean elevation within a predetermined neighbourhood. Using TPI, landscapes can be classified in slope position classes. Many researchers have used this index in the field of geomorphology (Tagil and Jenness, 2008; Liu et al., 2009; McGarigal et al., 2009), geology (Mora-Vallejo et al., 2008; Deumlich et al., 2010; Illés et al., 2011), hydrology (Lesschen et al., 2007; Francés and Lubczynski, 2011; Liu et al., 2011), agricultural science (Pracilio et al., 2006) and archaeology (Patterson, 2008; Berking et al., 2010).

The TPI is the difference of a cell elevation in a digital elevation model from the mean elevation ($\bar{X}$) of a user-specified neighbourhood surrounding. Local mean elevation is subtracted from the elevation value at centre of the local window (Gallant, 2000). The range of TPI depends not only on elevation differences but also on the adopted local window. Large local window values mainly reveal major landscape units, while smaller values highlight smaller features, such as minor valleys and ridges.

$$\mathrm{TPI}_i = X_0 - \bar{X}.$$

$$\bar{X} = \frac{\sum_{i=n} X_i}{n}.$$

where:

X_0 = elevation at the central point

$\bar{X}$ = average elevation around the central point within the local window

n = total number of surrounding points employed in the evaluation

TPI variation is shown in Figure 4.3 for arbitrary point elevations (A and B) in a DEM to the mean elevation of a specified neighbourhood around these point elevations. A small neighbourhood consisting of a 33×33 cell window was used to identify complex landscape features. The TPI provides a concise and effective technique of landscape classification in accordance with morphology.

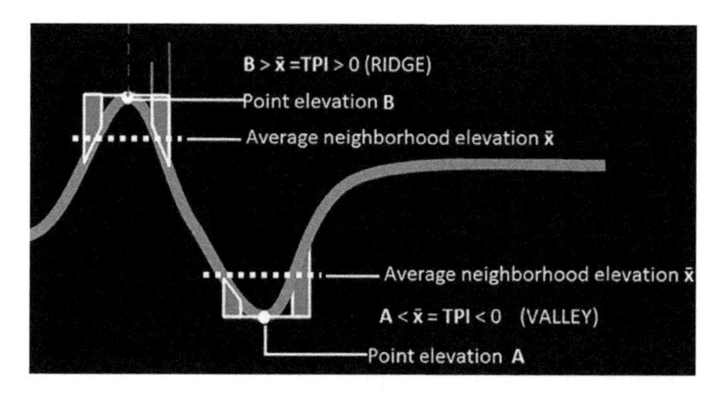

FIGURE 4.3
TPI variation is shown for arbitrary point elevations (A and B) in a DEM to the mean elevation of a specified neighbourhood around these point elevations.

TABLE 4.2
Recommend Standard Deviations (σ) Away from the Mean TPI Raster as Threshold Values for Classifying Six Slope Positions

Class	Description Breakpoints
Valley	$TPI \leq -1\sigma$
Lower slope	$-1\sigma < TPI \leq -0.5\,\sigma$
Flat slope	$-0.5\,\sigma < TPI < 0.5\sigma$, Slope $\leq 5^{\circ}$
Middle slope	$-0.5\,\sigma < TPI < 0.5\sigma$, Slope $\leq 5^{\circ}$
Upper slope	$0.5\,\sigma < TPI \leq 1\sigma$
Ridge	$TPI > 1\,\sigma$

More slope results in a higher runoff potential and low infiltration, and a lower degree of slope favours the retention of water and the drainage in the depth. There are a wide range of geomorphological methods and algorithms to classify the landscape into morphological classes (Burrough et al., 2000; Deng, 2007; Iwahashi and Pike, 2007; Hengl and Reuter, 2008). Weiss (2001) and Jenness (2006) recommend (Table 4.2) a standard deviation (σ) away from the mean TPI raster as threshold values for classifying six slope positions.

Land Use Land Cover

The land use and land cover or LULC patterns of given watershed influence the runoff. To compute hydrological elements more accurately, more accurate LULC maps are required. Using image processing techniques, images can be produced which depict some of the characteristics, notably the cover types such as areas with vegetation, water bodies, bare soils, etc. The LULC pattern and rainfall have a significant influence on the hydrological response of the watershed. A Sentinel-2 Level 1C data product (L1C_T43QCE_A008039_20180920T054434) acquired on 20 September 2018 was downloaded from the Sentinel Hub developed by the European Space Agency. Sentinel-2 Level 1C data were processed from Top-of-Atmosphere Level 1C to Bottom-of-Atmosphere Level 2A. The support vector machine classifiers were later used with principal components (PC) of the Sentinel-2 bands for classifying data into seven major LULC classes namely water, builtup, mixed forest, cultivated land, barren land, fallow land with Vertisols in dominance and fallow land with Inceptisols in dominance for the Vishwamitri watershed.

Principal component analysis (PCA) is a statistical procedure that transforms the input bands (with correlated variables) orthogonally from an input multivariate attribute space to a new multivariate attribute space having linearly uncorrelated variables whose axes are rotated with respect to each other. Transformation or dimensionality reduction of the data in PCA analysis compresses data by eliminating noise, redundancy and irrelevant information. The linearly uncorrelated variables in new multivariate attribute space are called principal components. The first principal component (PC1, derived from the first eigenvector) is the direction in space along which projections have the largest variance. The subsequent principal component (PC2) is the direction which maximizes variance among all directions orthogonal to the previous principal component. The variances of the remaining principal component images decrease in order, as denoted by the magnitudes of the corresponding eigenvalues Deng, 2007).

The accuracy assessment was done by comparing the real extent of the classes in the classified image relative to the reference data set using the error matrix, which is used to define the quality of the map derived from the data.

Accordingly, overall accuracy, producer's and user's accuracies and kappa coefficient were computed. The accuracy and quality of the reference data should be at least one order better as compared to the data to be evaluated. DigitalGlobe's WorldView-4 data (Product ID: 1ba34688-3ee0-41e4-9187-de68fdb075df-inv) acquired on 25 October 2018 at 5:30 am with 31 cm resolution was used for the accuracy assessment.

The error matrix is represented by a table that shows correspondence between the classification result and a reference image assigned to a particular category, which is relative to the actual category as indicated by the reference data. Producer's accuracy is the probability that value in each class was correctly classified (Rana and Suryanarayana, 2019a; b).

$$\text{Producer's accuracy} = \frac{\text{total number of correct pixels in a class}}{\text{total pixels in that class as derived from the reference data}}$$

User's accuracy is the probability that a value predicted to be in a certain class is really in that class.

$$\text{User's accuracy} = \frac{\text{total number of correct pixels in a class}}{\text{total pixels that were classified in that class}}$$

The kappa coefficient measures the agreement between classification and truth-values. A kappa value of 1 represents perfect agreement, while a value of 0 represents no agreement.

$$\text{Kappa coefficient} = \frac{\text{Observed accuracy} - \text{Expected agreement}}{1 - \text{Expected agreement}}$$

The overall accuracy is given by the ratio of the proportion of the correctly classified pixels to the total number of pixels in the confusion matrix.

Soil Texture

Soil texture refers to the relative proportions of clay, silt, and sand. Soils containing large proportions of sand have relatively large pores through which water can drain freely. These soils produce less runoff. As the proportion of clay increases, the size of the pore space decreases. This restricts movement of water through the soil and increases the runoff.

Soil data based on soil texture collected from National Bureau of Soil Survey and Land Use Planning (NBSS and LUP). The LULC maps were later used in Hydrologic Engineering Center's Geospatial Hydrologic Modeling Extension (HEC-GeoHMS) for the integration of LULC and soil data for Curve Number (CN) grid preparation. The HEC-GeoHMS is extension to ESRI's ArcGIS software that computes the CN and other loss rate parameters based on various soil and LULC databases.

Curve Number (CN)

The CN is the most used reliable and conceptual technique for estimating surface runoff. CN is basically a dimensionless number that reduces the rainfall to runoff. It depends upon two parameters LULC and Hydrologic Soil Group (HSG). HSG is one of the important parameters for assigning curve numbers and is generated by reclassifying the soil textural map considering their runoff potential into account (Singh et al., 2017; Rizeei et al., 2018; Tripathi, 2018; Hameed et al., 2019).

TABLE 4.3

Selected Curve Number Values for the Study Area Using TR-55 Table

LULC	Sub-Classes of LULC	HSG-A	HSG-B	HSG-C	HSG-D
Water	Water	100	100	100	100
Cultivated land	Cultivated land crop 1	64	75	82	85
	Cultivated land crop 2	71	80	87	90
Sparsely vegetated	Sparsely vegetated	74	83	88	90
Barren land	Barren land	77	86	91	94
Fallow land (Vertisols dominance)	Fallow land 1 Vertisols dominated	76	85	90	93
	Fallow land 2 Vertisols dominated	76	85	90	93
Fallow land (Inseptisol dominance)	Fallow land 1 Inseptisol dominated	74	83	88	90
	Fallow land 2 Inseptisol dominated	74	83	88	90

The CN grid is a raster containing CN values assigned to each grid cell of LULC and soil complex to indicate their specific runoff potential. A logical condition was defined in ArcGIS to generate the CN raster file from the raster files of the HSG and LULC using the TR-55 table (Feldman, 2000). CNs vary from 0 to 100 and express the runoff response to a given rainfall event. Higher CNs indicate a greater proportion of rainfall to be transformed into surface runoff. Selected CN values for the study area are given in Table 4.3.

Stream Order

The availability of the total quantity of surface water is proportional to the stream order and some structures are suitable at a particular drainage order only. For example, check dams should only be constructed at lower-order streams (IMSD, 1995; Durga and Bhaumik, 2003).

The stream order of the Vishwamitri watershed was assigned using the Strahler (1957) method, where all streams without any tributaries are assigned an order of 1 and are referred to as first order. The stream segments starting from the confluence of two streams of the first order are called streams of second order and so on. The tail point of each stream is defined as the point from where a stream of higher order starts. Flow accumulation and flow-direction rasters were used to generate stream networks using the hydrology toolset of ArcGIS. Stream ordering was done to assist in the proper planning of conservation measures in terms of storage and capacity.

Height Above Nearest Drainage (HAND)

The HAND is a digital elevation model normalized using the nearest drainage. It normalizes topography according to the local relative heights found along the drainage network and in this way presents the topology of the relative soil gravitational potential, or local draining potentials. HAND allows for the calculation of the elevation of each point in the catchment above the nearest stream it drains to, following the flow direction (Rennó et al., 2008; Nobre et al., 2011; Hamdani and Baali, 2019).

The HAND raster was prepared for the fourth- and fifth-order streams of the Vishwamitri watershed as they are highly susceptible to flooding. The first step was to remove small imperfection by filling sinks in the Cartosat-1 DEM. Sinks must be filled to ensure a proper delineation of streams. A derived drainage network may be discontinuous if the sinks are not filled (Rana and Suryanarayana, 2019a, b). The second step is to create a flow-direction raster and it is computed from the DEM using the D8 method (Jenson and Domingue, 1988) to determine the flow from each cell to its steepest downslope neighbour. An erroneous flow-direction raster may result in the presence of sinks. Next, the accumulated flow direction is used to find the nearest stream cell for each cell. At last, the elevation of the nearest stream cell is deducted from the elevation of each cell to normalize the terrain and to get its corresponding HAND value.

Determining Criteria Weights Using AHP

The AHP is one of the MCDM methods that was originally developed by Saaty (1987), and it has been widely applied to solve decision-making problems related to water resources. The approach combines mathematics and psychology in dealing with complex decisions and in turn converts it into a simpler system of hierarchy. This method compares two criteria at a time through a pairwise comparison matrix. Each criterion is assessed by arranging every possible pairing on a ratio scale to express the comparative importance by numerical values. The numerical expression of suitability rating (Burnside et al. 2002) and scaling of comparative importance (Saaty, 1990) are given in Table 4.4. The judgement on dominance of one criterion over another is based on the authors' expertise and a literature survey (Jha et al., 2014; Krois and Schulte, 2014; Prasad et al., 2014; Ammar et al., 2016; Bitterman et al., 2016; Singh et al., 2017; Wu et al., 2018).

The determination of the relative-importance weight of each criterion (Slope, TWI, LULC, Curve Number, Stream Order and HAND) for potential runoff storage zones was calculated by using the pairwise comparison matrix method. The number of comparisons can be determined using:

TABLE 4.4

Pairwise Comparison Scale for AHP Preferences

Numerical Expression	Suitability Rating	Comparative Importance
1	Not suitable	Equal importance
3	Marginally suitable	Moderate importance of one over another
5	Moderately suitable	Essential or strong importance
7	Highly suitable	Very strong importance
9	Optimally suitable	Extreme importance
2, 4, 6, 8	Intermediate values between the two adjacent judgements	
Reciprocal of above numbers	If one criterion has one of the above numbers assigned to it when compared with a second criterion, then the second criterion has the reciprocal value when compared to the first	

$$\text{Number of comparisons} = \frac{n(n-1)}{2}$$

where n = number of criteria.

The resulting pairwise comparison matrix is used to obtain the eigenvalue of each criterion, which represents its relative-importance weight (Saaty, 1990). The relative-importance weight given to the criteria one over another is acceptable if the consistency ratio (CR) is less than 10%. If it increases 10%, a new value is assigned in the pairwise comparison matrix. CR is computed as follows:

$$CR = \frac{\dfrac{(\lambda_{max} - n)}{(n-1)}}{RI}.$$

where λ_{max} is the principal eigenvalue, n is the number of elements compared and RI is the so-called random consistency index, a value that depends on the number of criteria that are being compared (Saaty, 1987).

Weighted Overlay Process (WOP) within GIS

After calculating weights for each criterion, the WOP was applied to construct suitability maps, also known as MCDM within the GIS environment. ArcGIS was used for the WOP computation where each criterion raster layer was assigned a calculated weight in the suitability analysis. Values in the rasters were reclassified to a common 1 (least suitable) to 9 (highly suitable) suitability scale. Each raster layer was multiplied by its weight and the results were summed according to the following equation (Malczewski, 1999):

$$A_j = \sum_{i=1}^{m} W_i X_{ij}.$$

where

A_j = final suitability score in each cell

X_{ij} = suitability of the ith cell with respect to the jth layer

W_i = normalised weight so that $\sum W_i = 1$

The resulted suitability map or potential runoff storage zones map was further classified into four classes: (1) not suitable, (2) marginally suitable, (3) moderately suitable and (4) optimally suitable.

Results and Discussion

Rainfall Analysis

Based on the past 56 years (1961–2016) of rainfall and data analysis of rain gauge stations across watershed, it was determined that the SPI indicated extremely dry years 1.8% of the time, moderately dry years 10.7% of the time, moderately wet years 3.6% of the time, near normal years 73.2% of the time, very wet years 7.1% of the time and extremely wet years 3.6% of the time. Classification of annual rainfall of the study area based on SPI is shown in Table 4.5.

The results therefore suggest that the overall drought events between these years were not severe. The precipitation analysis result suggests that the water shortage in the region is a management problem. This qualifies the area for identification of suitable sites for water storage. Also, computed annual rainfall for Vadodara, Savali, Padra and

TABLE 4.5

Classification of Annual Rainfall Based on SPI

Classification	Years
Dry extreme years	2008
Moderately dry years	1972, 1974, 1986, 1987, 1999, 2000
Moderately wet years	1983, 1994
Near normal years	1961, 1962, 1963, 1964, 1965, 1966, 1967, 1968, 1969, 1971, 1973, 1975, 1977, 1978, 1979, 1980, 1981, 1982, 1984, 1985, 1988, 1989, 1990, 1991, 1992, 1993, 1995, 1996, 1997, 1998, 2001, 2002, 2003, 2004, 2007, 2009, 2011, 2012, 2014, 2015, 2016
Very wet years	1976, 2006, 2010, 2013
Wet extreme years	1970, 2005

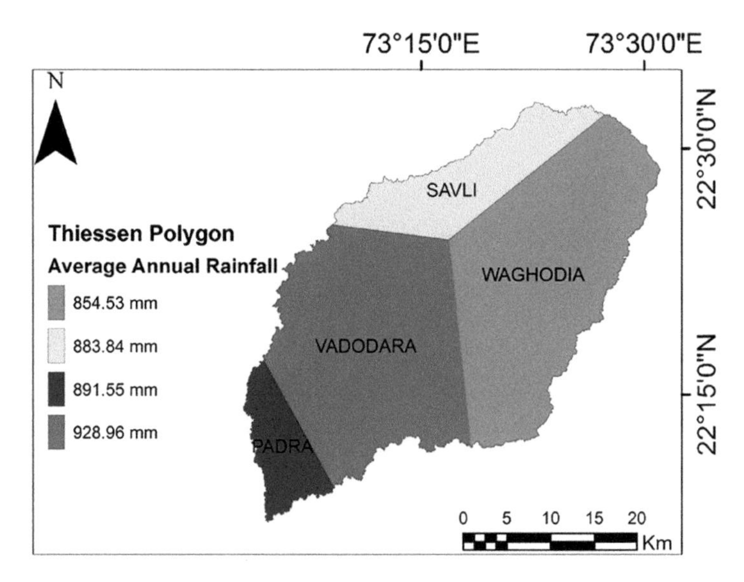

FIGURE 4.4
Thiessen polygon of rain gauge stations for Vishwamitri watershed.

Waghodia rain gauge stations were 928.96, 891.56, 883.85 and 854.53 mm, respectively.

The seasonal distribution of the precipitation in the study area varies and falls mostly as rain in monsoon seasons (June to September). Thiessen polygons of rain gauge stations are shown in Figure 4.4. Promoting rainwater harvesting in areas receiving less than 100 mm/year or more than 1,000 mm/year of rains is not recommended (Mou et al., 1999; FAO, 2003; Mati et al., 2006). Water-based activities are not feasible in areas that receive less than 100 mm/year of rain, and there is no incentive to implement rainwater-harvesting schemes in areas with annual rains more than 1,000 mm/year. Results of rainfall analysis of the Vishwamitri watershed show potential to successfully carry out water-based activities.

Topography Wetness Index (TWI)

High values of the TWI were found in converging and flat areas and are expected to have much water accumulation and low slope. In contrast, steep locations and diverging areas received a small index value and have relatively lower water accumulation. Consequently, the index is a relative measure of the hydrological conditions of a given location in the landscape (Figure 4.5).

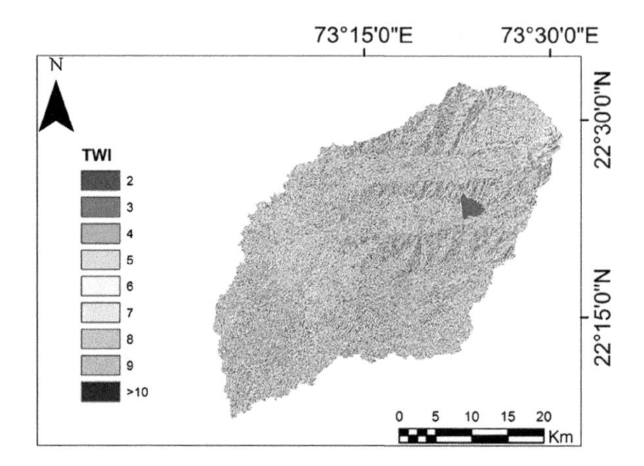

FIGURE 4.5
The calculated TWI for the Vishwamitri watershed.

Generation of Slope Maps Using Topography Position Index (TPI)

Positive TPI values indicate that the target point location is higher than the average of its surroundings, as defined by the neighbourhood (ridges). Figure 4.6 shows conversion of elevation values to TPI along the cross section (shown in red colour in TPI map). Negative TPI values represent locations that are lower than their surroundings (valleys). TPI values near zero are either flat areas (where the slope is near zero) or areas of constant slope (where the slope of the point is significantly greater than zero). TPI for the study area is shown in Figure 4.7.

Slope classification was done as suggested by Weiss (2001) and Jenness (2006). Slope plays a significant role in the amounts of runoff and sedimentation, the speed of water flow and the amount of material required to construct a dyke (the required height). Results show (Table 4.6 and Figure 4.8) that maximum area belongs to the flat class (35.45%), and flat areas are never strictly horizontal. Also, flat slopes lead to a decrease in the surface runoff velocity, which results in a longer period for the runoff to drain; there are gentle slopes in a seemingly flat area. Ponds are suitable for small flat areas with slopes 5%, where 0.15% belongs to middle slope. Nala bunds are suitable on moderate slopes of 5%–10%, and the 12.49% area belongs to upper slope. Terracing is suitable for steeper slopes of 5%–30%.

Ridges and upper slopes together form 30.79% of the area, where they indicate the least potential for rainwater harvesting because higher sloping land is inappropriate for constructing water storage structures. Valley and lower slope areas together constitute 33.58% of the area, and small dams or check dams like structures are preferable on such sites.

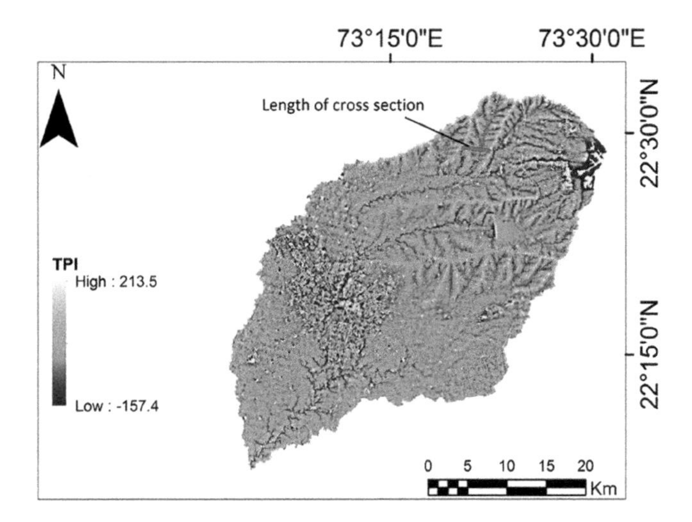

FIGURE 4.6
Comparison between the original DEM and TPI along the cross-sectional profile (red line).

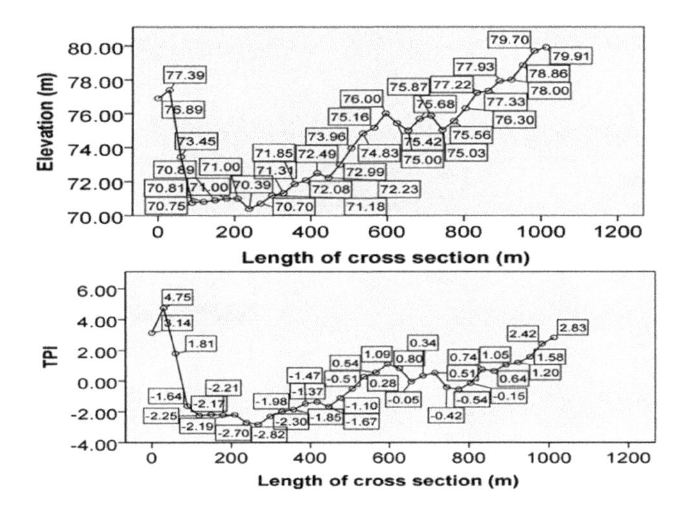

FIGURE 4.7
Variation of topography position index across the Vishwamitri watershed.

Land Use/Land Cover

LULC classes were selected based on knowledge about the specific study area. Seven major LULC classes were identified, viz. Water, Builtup, Mixed forest, Cultivated land, Barren land, Fallow land with Vertisols in dominance and Fallow land with Inceptisols in dominance. Figure 4.9a shows the three major PCA bands accounting for 97.96% of the eigenvalues.

TABLE 4.6

Slope Classifications Using TPI as the Basis of Landform Classification for the Study Area

Class	Description Breakpoints	Area (km²)
Valley	TPI≤–4.6	255
Lower slope	–4.6< TPI≤–2.3	178
Flat slope	–2.3<TPI<2.3, Slope≤5°	457
Middle slope	–2.3<TPI<2.3, Slope>5°	2
Upper slope	2.3<TPI≤4.6	161
Ridge	TPI>4.6	236

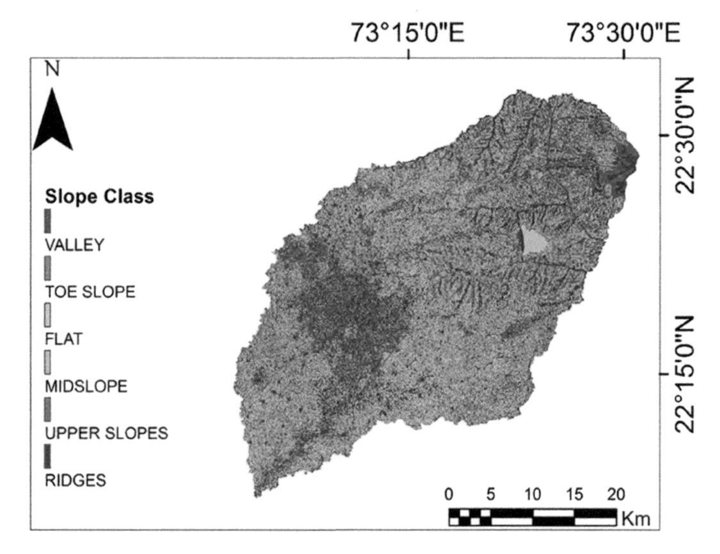

FIGURE 4.8
Resulted slope map of the study area using TPI as basis of landform classification.

The classification was conducted using Sentinel–2/PCA-based approach. Prediction performances of the Support Vector Machine were evaluated with training data of less than 1,000 pixels per class. Training data for each LULC class were collected as a group of pixels. Stratified random sampling was used to obtain the testing data. Figure 4.9b shows the variation of responses of the LULC classes in PCA-based approach.

Response Curves of LULC Classes in PCA Bands

The spectral response curve is the curve showing the variation of reflectance or emittance (in terms of digital numbers) of a material with respect to

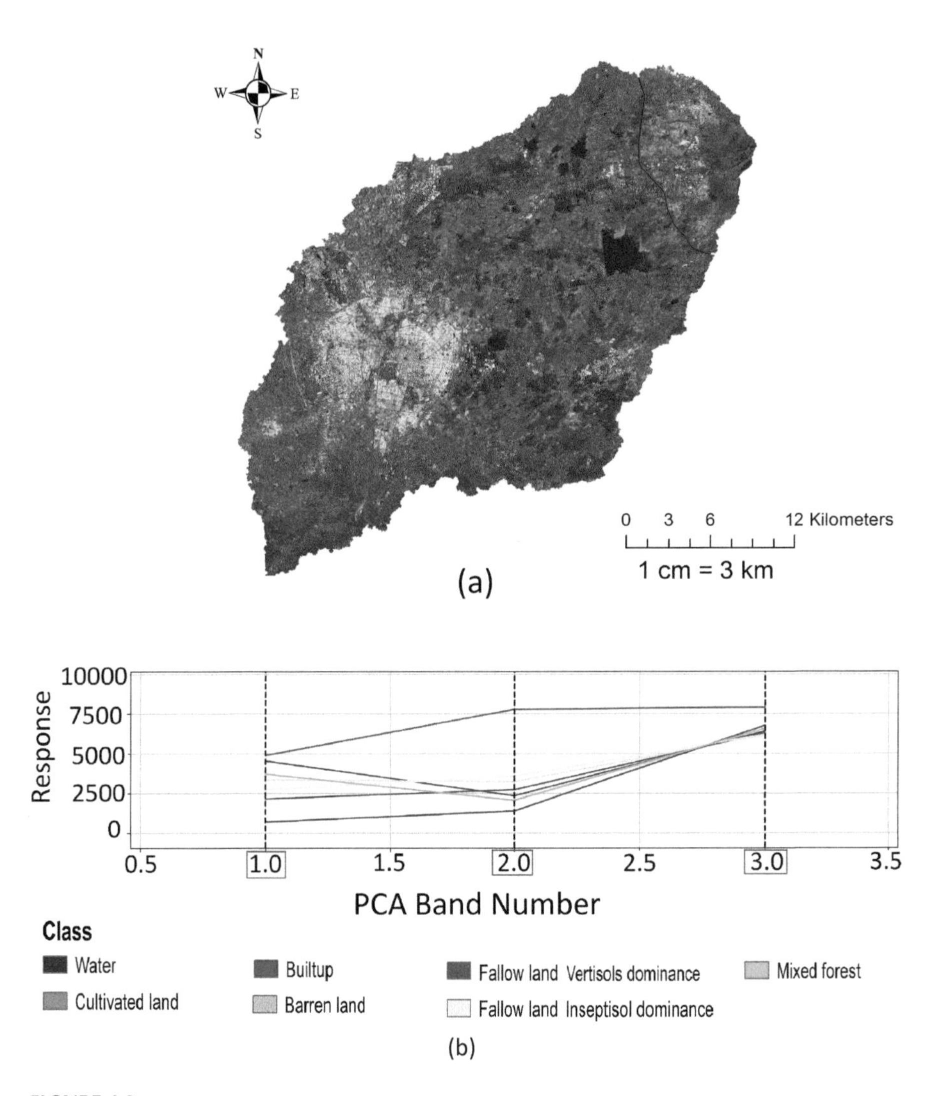

FIGURE 4.9
(a) The first three principal components visualized in false colour composites (b).

wavelengths. Classes having similar responses were hard to separate. It was also observed that the spectral distance (or separability) of classes Water and Builtup in relation to other classes was more so, that is why the user's accuracy (UA) and producer's accuracy (PA) for the water class were the highest followed by the Builtup class, for the PCA-based classification approach. Kappa coefficient showed a similar trend as that of the overall accuracy.

The derived LULC map of the study area is shown in Figure 4.10, which reveals that there are seven major types of land use/land cover namely Waterbodies, Builtup, Mixed forest, Cultivated land, Barren land, Fallow land with Vertisols

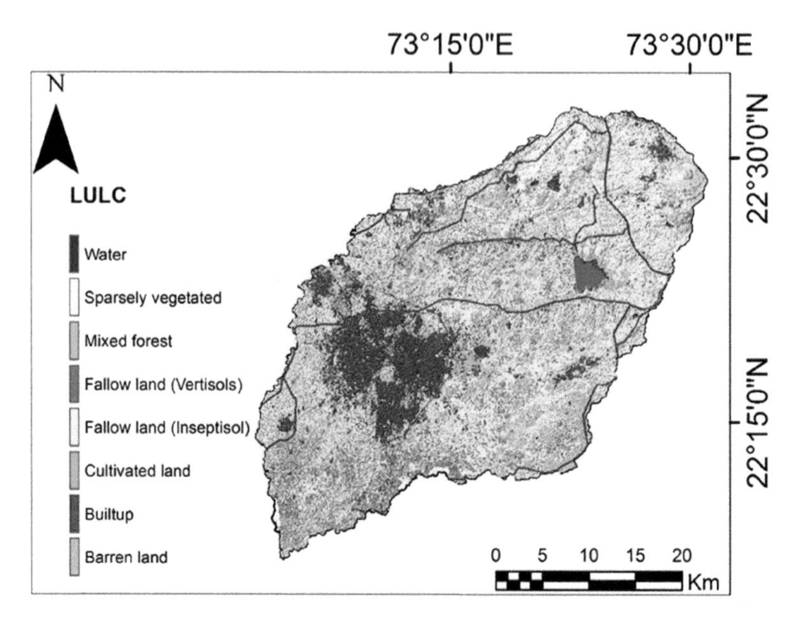

FIGURE 4.10
Land use land cover (LULC) map of the Vishwamitri watershed.

in dominance and Fallow land with Inceptisols in dominance. The major portion of the study area (about 35%) was Agricultural land (cultivated land, fallow land with Vertisols in dominance and fallow land with Inseptisols in dominance) followed by Sparsely vegetated (19%), Mixed forest (14%), Builtup (12%), Barren land (5%) and Water bodies (2%). LULC classes such as barren land and sparsely vegetated land are generally recommended for water storage zones/ structures. The results of accuracy assessment are given in Table 4.7.

Soil Texture

According to the classification system based on soil texture, seven types of soils are found in the Vishwamitri watershed (Figure 4.11). Typic Ustifluvents and Fluventic Haplustepts correspond to HSG group A, Udic Haplustepts correspond to HSG group B, Chromic Haplusterts and Typic Haplustepts correspond to HSG group C, and Lithic Haplustepts and Vertic Haplustepts correspond to the HSG group D. The most dominating soil, Chromic Haplusterts (HSG group C), covers 48.94% of the total watershed area. HSG-A has the lowest runoff potential (typically containing more than 90% sand and less than 10% clay), HSG-B has moderately low runoff potential (typically contains between 10% and 20% clay and 50%– and 90% sand), HSG-C has moderately high

TABLE 4.7

Accuracy Assessment Results

LULC	Sub-Classes of LULC	Producer's Accuracy	User's Accuracy
Water	Water	1.00	1.00
Cultivated land	Cultivated land Crop 1	0.80	0.57
	Cultivated land Crop 2	0.60	0.67
	Sparsely vegetated	0.88	0.88
Barren land	Barren land	0.50	0.43
Fallow land (Vertisols dominance)	Fallow land 1 Vertisols dominance	0.60	0.86
	Fallow land 2 Vertisols dominance	1.00	0.60
Fallow land (Inseptisol dominance)	Fallow land 1 Inseptisol dominance	1.00	0.67
	Fallow land 2 Inseptisol dominance	0.75	1.00
Mixed forest	Mixed forest	0.33	0.25
Builtup	Builtup	1.00	0.86
	Mixed Builtup 1	1.00	1.00
	Mixed Builtup 2	0.56	1.00
	Kappa coefficient	0.74	
	Overall Accuracy (%)	76.00	

runoff potential (typically contains between 20% and 40% clay and less than 50% sand) and HSG-D has high runoff potential (typically containing more than 40% clay and less than 50% sand) (Ross et al., 2018).

Curve Number (CN)

CNs consider the relationship between land use/land cover and HSGs, which together make up the curve number. The CN value varied from 36 to 100 (Figure 4.12) with a mean value of 82.39 and standard deviation of 11.56, where smaller numbers indicate low runoff potential, while larger numbers indicate an increased runoff potential. A CN value of 100 represents a condition of zero potential maximum retention that suggests an impermeable catchment having maximum runoff-generation capability. On the other hand, a CN value of 0 suggests an infinitely abstracting catchment having zero runoff-generation capability (Jha et al., 2014).

With this approach, the suitable locations for surface water storage zones/structures were in areas with the highest capacity for runoff generation and nearby to existing drainage lines. Several researchers have applied the CN method to identify potential runoff-harvesting sites (e.g., De Winnaar et al., 2007; Jha et al., 2014; Krois and Schulte, 2014).

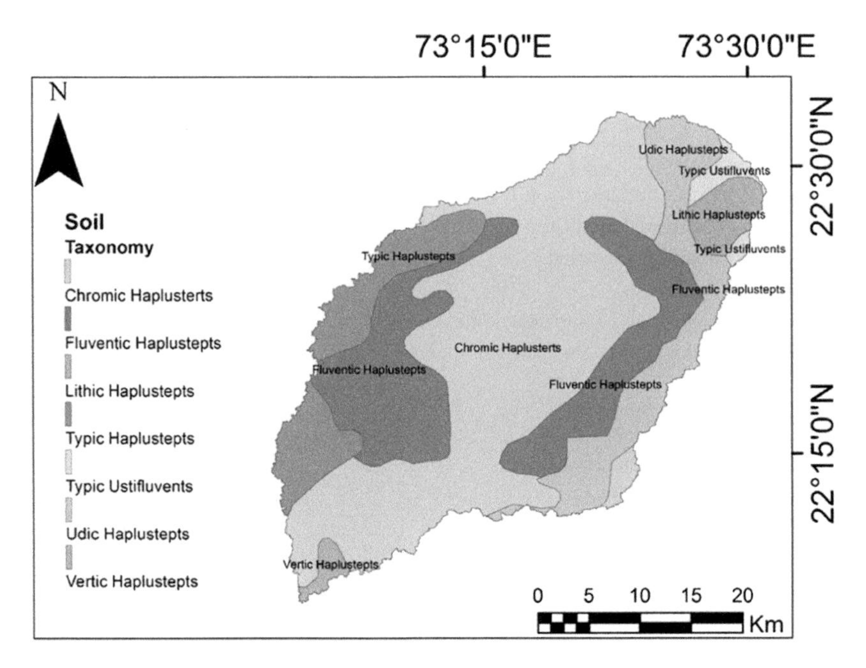

FIGURE 4.11
Soil data based on soil texture collected from (NBSS and LUP).

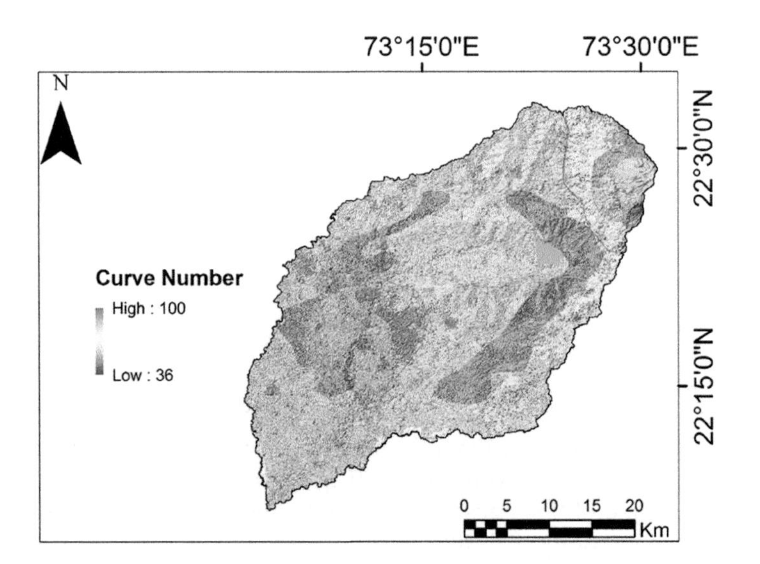

FIGURE 4.12
Variation of curve number across the Vishwamitri watershed.

Stream Order

All the first-, second- and third-order streams were extracted from the drainage network map and a stream-order buffer map was developed with a buffer distance of 50 m on both sides of the streams. It can be seen from the drainage network map (Figure 4.13) that the study area has a fifth-order drainage network, with a good drainage network in the eastern portion. The length of the first-order streams in the study area was nearly 294.5 km (about 52.9% of total length of drainage). The second- and third-order streams also exhibited considerable drainage lengths of 124.7 km (22.4%) and 88.1 km (15.8%), respectively. The fourth-order streams contributed to the drainage with a length of 33.0 km and accounting for 5.9% of the total drainage length. Mainly, the Vishwamitri river is a fifth-order stream having a drainage length of 16.7 km (3.0%).

Height Above Nearest Drainage (HAND)

Low-lying lands adjacent to streams are more susceptible to be flooded than higher land. HAND values for the study area varied from 0 to 749 m. HAND raster was prepared for the fourth- and fifth-order streams of the Vishwamitri watershed as they are highly susceptible to flooding. On the suitability scale, lower values were assigned for HAND values ranging from 0 to 2 m as these areas are more prone to flooding and it is not recommended to build water

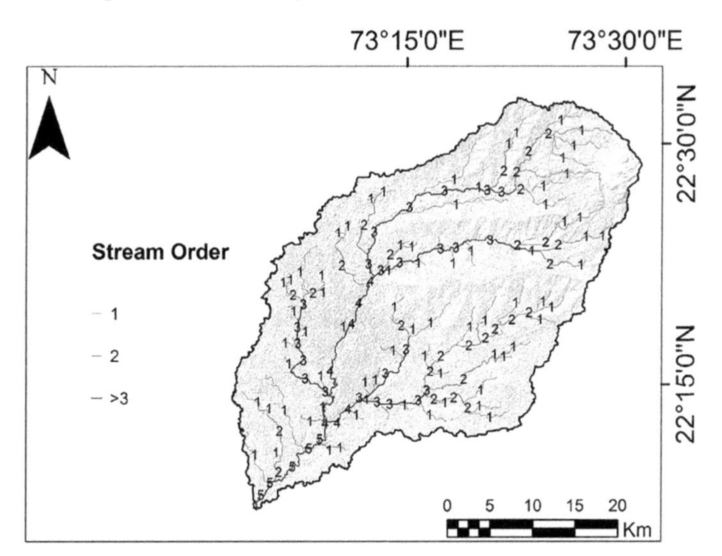

FIGURE 4.13
Drainage network map showing stream order of the Vishwamitri watershed.

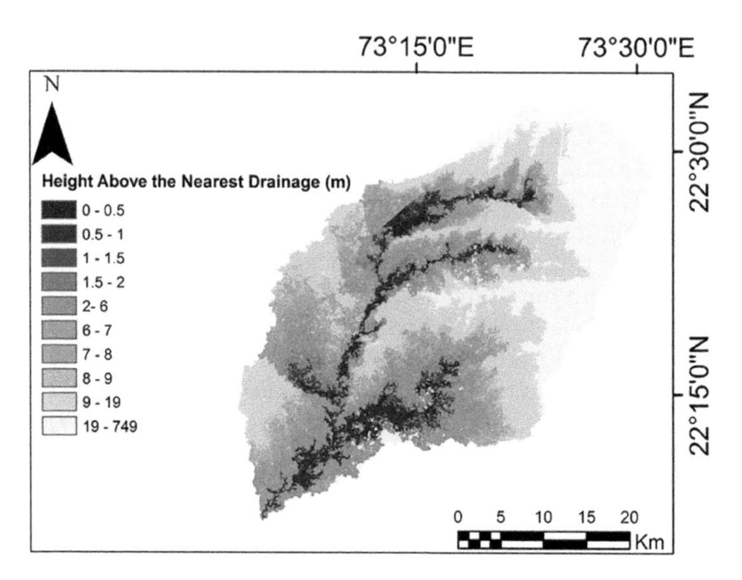

FIGURE 4.14
Height above nearest drainage (HAND) map of the Vishwamitri watershed.

storage structure on such zones. Also, lower values were assigned on the suitability scale for extremely high HAND values (greater than 48 m) because higher HAND value means points move away from the river. Figure 4.14 shows the Height Above Nearest Drainage map of the study area.

Determining Criteria Weights Using AHP

AHP provided a systematic approach to conduct MCDM. To derive suitability maps for potential runoff storage zones, the criteria maps must be related to the result of the AHP. The AHP pairwise matrix for the criteria used in this study is presented in Table 4.8. For all the five spatial layers the relative importance was derived. Thus, a common scale (0%–100%) was obtained from the AHP procedure. As seen in Table 4.8, the most important criterion for decision-making was Slope (22.50%), followed by LULC (21.20%), CN (14.80%), HAND (14.50%), Stream order (14.40%) and TWI (12.60%). The calculated principal eigenvector was 6.47, which was computed with the square reciprocal matrix of pairwise comparisons between criteria. Since the AHP may have inconsistencies in establishing the values for the pairwise comparison matrix, it is important to calculate this level of inconsistency using the consistency ratio (CR). The CR of the pairwise matrix was 7.5% (which is less than 10%) and thus the

TABLE 4.8
Resulting Weights for the Criteria Based on Pairwise Comparisons

	Slope	TWI	LULC	Curve Number	Stream Order	HAND	Priority	Rank
Slope	1	2	0.5	2	2	2	22.50%	1
TWI	0.5	1	1	1	1	0.5	12.60%	6
LULC	2	1	1	1	2	1	21.20%	2
Curve number	0.5	1	1	1	0.5	2	14.80%	3
Stream order	0.5	1	0.5	2	1	1	14.40%	5
HAND	0.5	2	1	0.5	1	1	14.50%	4

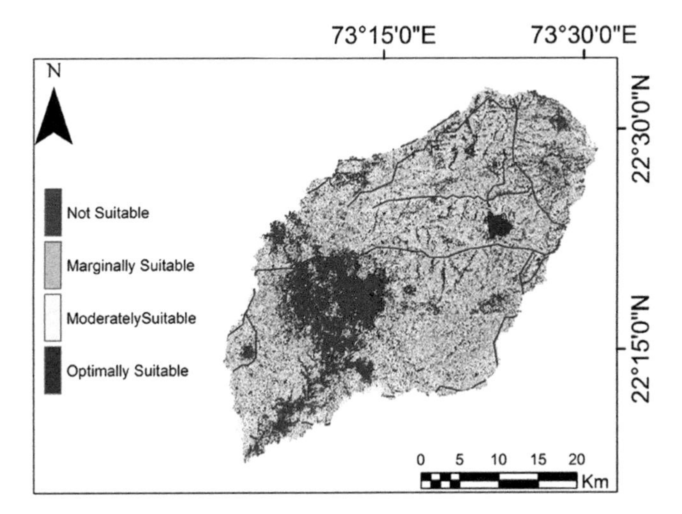

FIGURE 4.15
Potential runoff storage zones of the study area.

judgements made and compiled in the pairwise matrix of Table 4.8 were acceptable. This implies that the comparisons were performed with good judgement, and weightage for each criterion was suitable to weighted overlay.

Weighted Overlay Process (WOP) within the GIS

Potential runoff storage zones (Figure 4.15) were generated by integrating the thematic layers of slope, LULC, Curve Number, HAND, stream order and TWI using WOP within GIS. Resulting rasters were classified into four classes namely: (1) not suitable, (2) marginally suitable, (3) moderately suitable or (4) optimally suitable. Results showed that 17% of the area was optimally suitable, 33.2% of the area was moderately suitable, 33.1% of the area was marginally suitable and 18.7% of the area was not suitable for water storage zones/structures. Sixteen suitable sites on such zones (optimally suitable class) have also been identified for water storage structures, as shown in Figure 4.16.

Criteria of selection for these sites were first, proximity of the sites to the agricultural fields; second, sites should be on unused or barren land; and third, the site should be in the narrow cross section of the valley with high shoulders to minimize the amount of construction material needed for building the small dams, check dams, nala bunds, gully plugs and bundhis. Results were also confirmed by the already built water storage structures in derived potential runoff storage zones which are in optimally suitable class (Figure 4.17).

Conclusions

The objective was to identify potential runoff storage zones based on the various physical characteristics of the Vishwamitri watershed using a GIS-based concept. The conceptual framework combines factors through an AHP using the MCDM method. Further, the selection of factors (Slope, Land use/land cover, HAND, Stream order, Curve number, TWI) was based on review suggestions from several previous investigators. Results will help concerned authorities in the proficient arranging and execution of water-related plans and schemes, improve water shortage, reduce dependability on ground water and ensure sustainable water availability for local and agricultural purposes. It was recommended that the proposed suitability map developed by the GIS techniques be implemented to help overcome future growing water scarcity due to global/regional climate change. The approach and analysis showed in this approach can be generalized to other places and potentially being of relevance and valuable for application in other parts of the world. This may be especially true for developing countries, despite hydrological and agro-climatic variations. This approach can be potentially less time-consuming, more precise and can be utilized for identifying locations for different interventions for large watersheds.

Acknowledgement

Part of the contribution's contents and figures are used with permission by Taylor & Francis Group and from "Rana, V. K., and Suryanarayana,

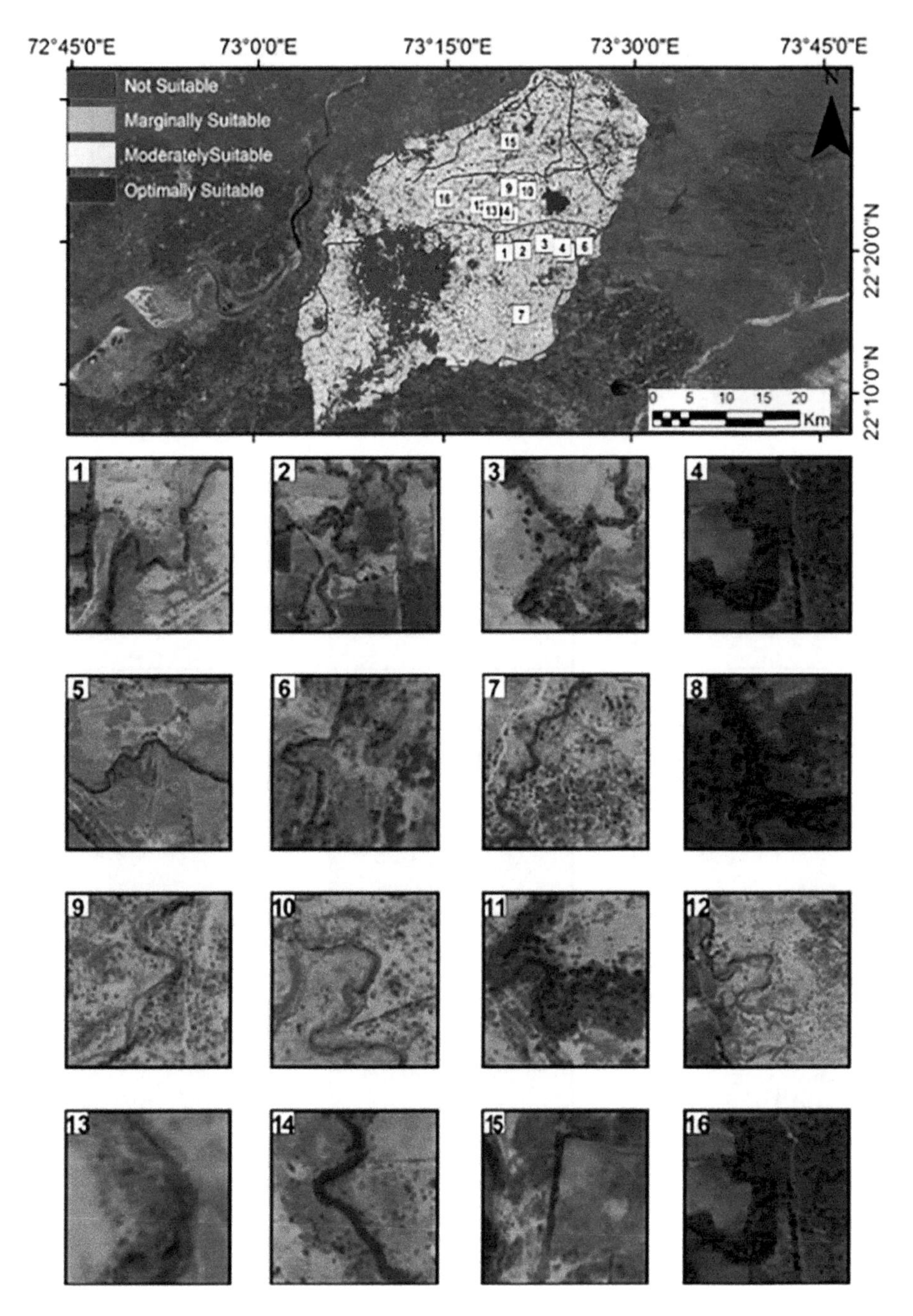

FIGURE 4.16
Identified sites for water storage structures on potential runoff storage zones.

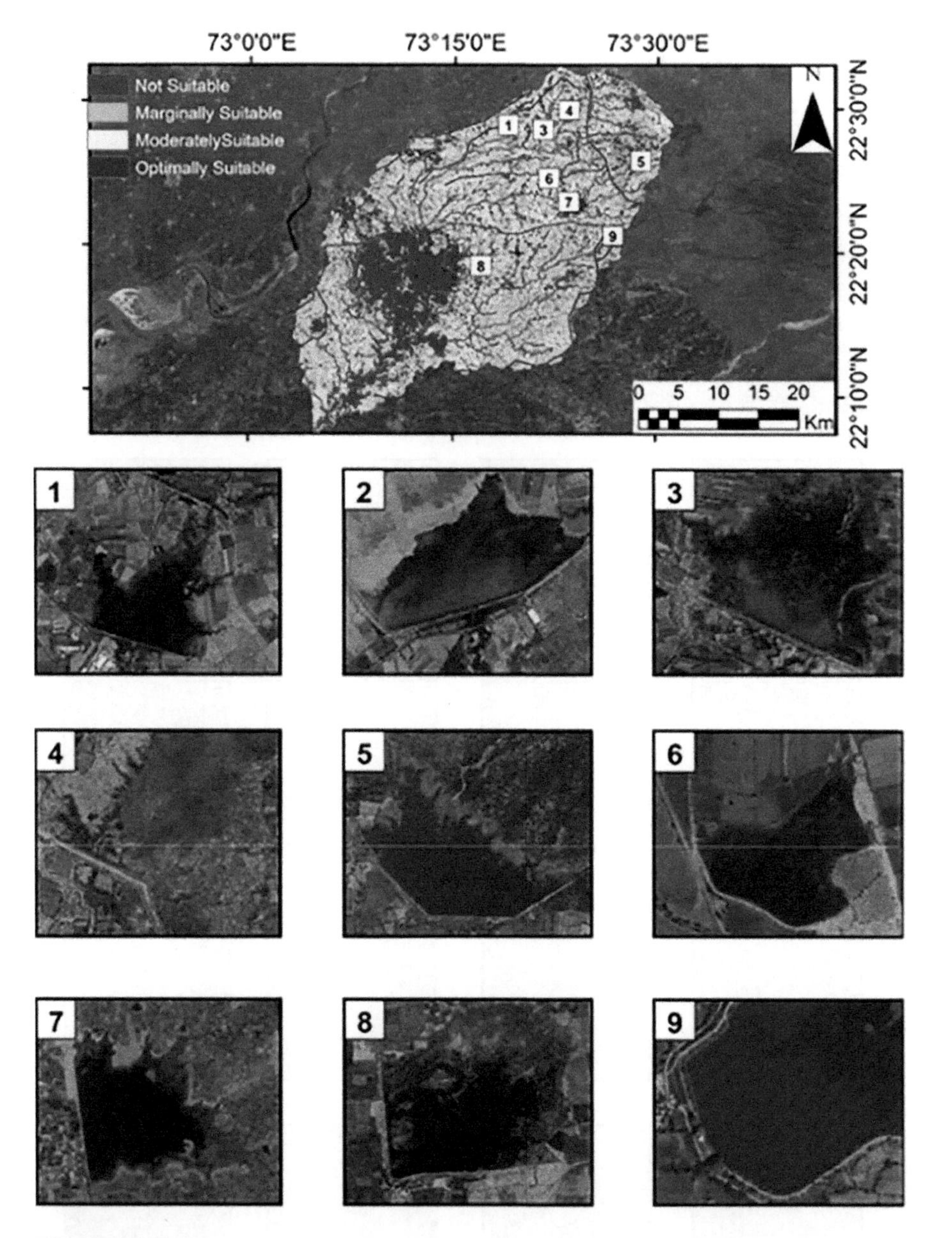

FIGURE 4.17
Already built water storage structures on the derived potential runoff storage zones.

T. M. V. 2020. GIS-based multi criteria decision-making method to identify potential runoff storage zones within watershed. *Annals of GIS* 26(2):149–168."

References

Adegoke, C. W., and Sojobi, A. O. 2015. Climate change impact on infrastructure in Osogbo metropolis, south-west Nigeria. *Journal of Emerging Trends in Engineering and Applied Sciences*, 6, 156–165.

Ågren, A. M., Lidberg, W., Strömgren, M., Ogilvie, J., and Arp, P. A. 2014. Evaluating digital terrain indices for soil wetness mapping–A Swedish case study. *Hydrology and Earth System Sciences*, 18(9), 3623–3634.

Akinsanola, A. A., and Ogunjobi, K. O. 2014. Analysis of rainfall and temperature variability over Nigeria. *Global Journal of Human Social Sciences: Geography and Environmental GeoSciences*, 14(3), 1–18

Al-Adamat, R. 2008. GIS as a decision support system for siting water harvesting ponds in the Basalt Aquifer/NE Jordan. *Journal of Environmental Assessment Policy and Management*, 10(2), 189–206.

Ammar, A., Riksen, M., Ouessar, M., and Ritsema, C. 2016. Identification of suitable sites for rainwater harvesting structures in arid and semi-arid regions: A review. *International Soil and Water Conservation Research*, 4(2), 108–120.

Berking, J., Beckers, B., and Schütt, B., 2010. Runoff in two semi-arid watersheds in a geoarcheological context: A case study of Naga, Sudan, and Resafa, Syria. *Geoarchaeology* 25, 815–836.

Beven, K.J., and Kirkby, M.J., 1979. A physically based, variable contributing area model of basin hydrology. *Hydrological Sciences Bulletin* 24 (1), 43–69.

Bitterman, P., Tate, E., Van Meter, K. J., and Basu, N. B. 2016. Water security and rainwater harvesting: A conceptual framework and candidate indicators. *Applied Geography*, 76, 75–84.

Bjelanovic, I. 2016. Predicting forest productivity using Wet Areas Mapping and other remote sensed environmental data. Master of Science Thesis, Department of Renewable Resources, University of Alberta, Edmonton.

Burnside, N., Smith, R., and Waite, S. 2002. Habitat suitability modelling for calcareous grassland restoration on the South Downs, United Kingdom. *Journal of Environmental Management*, 65(2), 209–221.

Burrough, P. A., van Gaans, P. F., and MacMillan, R. A. 2000. High-resolution landform classification using fuzzy k-means. *Fuzzy Sets and Systems*, 113(1), 37–52.

de Castro-Pardo, M., Fernández Martínez, P., Pérez Zabaleta, A., and Azevedo, J. C. 2021. Dealing with water conflicts: A comprehensive review of MCDM approaches to manage freshwater ecosystem services. *Land*, 10(5), 469.

De Winnaar, G., Jewitt, G. P. W., and Horan, M. 2007. A GIS-based approach for identifying potential runoff harvesting sites in the Thukela River basin, South Africa. *Physics and Chemistry of the Earth*, 32, 1058–1067.

Deng, Y. 2007. New trends in digital terrain analysis: Landform definition, representation, and classification. *Progress in Physical Geography*, 31(4), 405–419.

Deumlich, D., Schmidt, R., and Sommer, M., 2010. A multiscale soil–landform relationship in the glacial-drift area based on digital terrain analysis and soil attributes. *Journal of Plant Nutrition and Soil Science*, 173, 843–851.

Dilts, T.E., 2015. *Topography Tools for ArcGIS 10.1*. ESRI, Redlands, CA.

Durga Rao, K. H. V., and Bhaumik, M. K. 2003. Spatial expert support system in selecting suitable sites for water harvesting structures—A case study of song watershed, Uttaranchal, India. *Geocarto International*, 18(4), 43–50.

Eshghizadeh, M., Talebi, A., and Dastorani, M. T. 2018. A modified lapsus model to enhance the effective rainfall estimation by SCS-CN method. *Water Resources Management*, 32(10), 3473–3487.

FAO, 2003. Land and Water Digital Media Series, 26. Training Course on RWH (CDROM). Planning of Water Harvesting Schemes, Unit 22. Food and Agriculture Organization of the United Nations, Rome, FAO.

Feldman, A. D. 2000. *Hydrologic Modeling System HEC-HMS: Technical Reference Manual*. US Army Corps of Engineers, Hydrologic Engineering Center, Davis, CA.

Fischer, G., F. N. Tubiello, H. van Velthuizen, and D. A. Wiberg. 2007. Climate change impacts on irrigation water requirements: Effects of mitigation, 1990–2080. *Technological Forecasting and Social Change*, 74, 1083–1107.

Francés, A.P., and Lubczynski, M.W., 2011. Topsoil thickness prediction at the catchment scale by integration of invasive sampling, surface geophysics, remote sensing and statistical modeling. *Journal of Hydrology*, 405, 31–47.

Gallant, J.C., and Wilson, J.P., 2000. Primary topographic attributes. In: Wilson, J.P., Gallant, J.C. (Eds.), *Terrain Analysis: Principles and Applications*. Wiley, New York, pp. 51–85.

Ghani, M. W., Arshad, M., Shabbir, A., Mehmood, N., and Ahmad, I. 2013. Investigation of potential water harvesting sites at Potohar using modeling approach. *Pakistan Journal of Agricultural Sciences*, 50(4), 723–729.

Hamdani, N., and Baali, A. 2019. Height Above Nearest Drainage (HAND) model coupled with lineament mapping for delineating groundwater potential areas (GPA). *Groundwater for Sustainable Development*, 9, 100256.

Hameed, H. M., Faqe, G. R., and Rasul, A. 2019. Effects of land cover change on surface runoff using GIS and remote sensing. In *Environmental Remote Sensing and GIS in Iraq*, Springer, Berlin, Germany, p. 205.

Hengl, T., and Reuter, H. I. (Eds.). 2008. *Geomorphometry: Concepts, Software, Applications*. Developments in Soil Science, vol. 33, Elsevier, Dordrecht, NL, p. 772.

Hojati, M., and Mokarram, M. 2016. Determination of a topographic wetness index using high resolution digital elevation models. *European Journal of Geography*, 7, 41–52.

Illés, G., Kovács, G., and Heil, B., 2011. Comparing and evaluating digital soil mapping methods in a Hungarian forest reserve. *Canadian Journal of Soil Science* 91, 615–626.

IMSD, 1995. *Integrated Mission for Sustainable Development Technical Guidelines*, National Remote Sensing Agency, Department of Space, Government of India, Hyderabad.

Iwahashi, J., and Pike, R. J. 2007. Automated classifications of topography from DEMs by an unsupervised nested-means algorithm and a three-part geometric signature. *Geomorphology*, 86(3–4), 409–440.

IWMI. 2007. Comprehensive assessment of water management in agriculture. In: *Water for Food, Water for Life: A Comprehensive Assessment of Water Management in Agriculture*. International Water Management Institute, London and Colombo.

Jenness, J. 2006. *Topographic Position Index (tpi_jen.avx) Extension for ArcView 3.x, v.1.3a.* ESRI Press, Redlands, CA.
Jenson, S. K., and Domingue, J. O. 1988. Extracting topographic structure from digital elevation data for geographic information system analysis. *Photogrammetric Engineering and Remote Sensing*, 54(11), 1593–1600.
Jha, M. K., Chowdary, V. M., Kulkarni, Y., and Mal, B. C. 2014. Rainwater harvesting planning using geospatial techniques and multicriteria decision analysis. *Resources, Conservation and Recycling*, 83, 96–111.
Kahinda, J. Mwenge, E. S. B. Lillie, A. E. Taigbenu, M. Taute, and R. J. Boroto. 2008. Developing suitability maps for rainwater harvesting in South Africa. *Physics and Chemistry of the Earth, Parts A/B/C*, 33(8–13), 788–799.
Karimi, H., and Zeinivand, H. 2019. Integrating runoff map of a spatially distributed model and thematic layers for identifying potential rainwater harvesting suitability sites using GIS techniques. *Geocarto International*, 36(3), 320–339.
Krois, J., and Schulte, A. 2014. GIS-based multi-criteria evaluation to identify potential sites for soil and water conservation techniques in the Ronquillo watershed, northern Peru. *Applied Geography*, 51, 131–142.
Kumar, T., and Jhariya, D. C. 2017. Identification of rainwater harvesting sites using SCS-CN methodology, remote sensing and Geographical Information System techniques. *Geocarto International*, 32(12), 1367–1388.
Lesschen, J.P., Kok, K., Verburg, P.H., and Cammeraat, L.H., 2007. Identification of vulnerable areas for gully erosion under different scenarios of land abandonment in southeast Spain. *Catena*, 71, 110–121.
Liu, H., Bu, R., Liu, J., Leng, W., Hu, Y., Yang, L., and Liu, H., 2011. Predicting the wetland distributions under climate warming in the Great Xing'an Mountains, northeastern China. *Ecological Research*, 26, 605–613.
Liu, M., Hu, Y., Chang, Y., He, X., and Zhang, W., 2009. Land use and land cover change analysis and prediction in the upper reaches of the Minjiang River, China. *Environmental Management*, 43, 899–907.
Loritz, R., Kleidon, A., Jackisch, C., Westhoff, M., Ehret, U., Gupta, H., and Zehe, E. 2019. A topographic index explaining hydrological similarity by accounting for the joint controls of runoff formation. *Hydrology and Earth System Sciences*, 23(9), 3807–3821.
Mahmoud, S. H., and Alazba, A. A. 2014. The potential of in situ rainwater harvesting in arid regions: Developing a methodology to identify suitable areas using GIS based decision support system. *Arabian Journal of Geosciences*, 8, 5167–5179.
Malczewski, J., 1999. *GIS and Multicriteria Decision Analysis.* John Wiley & Sons, Inc., New York, p. 392.
Mati, B., De Bock, T., Malesu, M., Khaka, E., Oduor, A., Meshack, M., and Oduor, V., 2006. Mapping the Potential of Rainwater Harvesting Technologies in Africa. A GIS Overview on Development Domains for the Continent and Ten Selected Countries. Technical Manual No. 6. World Agroforestry Centre (ICRAF), Netherlands Ministry of Foreign Affairs, Nairobi, Kenya, 126 pp.
McGarigal, K., Tagil, S., and Cushman, S., 2009. Surface metrics: An alternative to patch metrics for the quantification of landscape structure. *Landscape Ecology*, 24, 433–450.
McKee, T.B., N.J. Doesken, and J. Kleist, 1993. The relationship of drought frequency and duration to time scale. In: *Proceedings of the Eighth Conference on Applied Climatology, Anaheim, California.* American Meteorological Society, Boston, MA, pp. 179–184.

Mora-Vallejo, A., Claessens, L., Stoorvogel, J., and Heuvelink, G.B.M., 2008. Small scale digital soil mapping in southeastern Kenya. *Catena*, 76, 44–53.

Mou, H., Wang, H., and Kung, H., 1999. Division study of rainwater utilization in China. In: *9th International Rainwater Catchment System Conference, Brazil.*

Nobre, A. D., Cuartas, L. A., Hodnett, M., Rennó, C. D., Rodrigues, G., Silveira, A., Waterloo, M. and Saleska, S. 2011. Height Above the Nearest Drainage–a hydrologically relevant new terrain model. *Journal of Hydrology*, 404(1–2), 13–29.

Patterson, J.J., 2008. Late Holocene land use in the Nutzotin Mountains: Lithic scatters, viewsheds, and resource distribution. *Arctic Anthropology*, 45, 114–127.

Pauw, E. D., Oweis, T., and Youssef, J. 2008. *Integrating Expert Knowledge in GIS to Locate Biophysical Potential for Water Harvesting: Methodology and a Case Study for Syria.* ICARDA, Aleppo, Syria.

Pracilio, G., Smettem, K., Bennett, D., Harper, R., and Adams, M., 2006. Site assessment of a woody crop where a shallow hardpan soil layer constrained plant growth. *Plant and Soil* 288, 113–125.

Prasad, H. C., Bhalla, P., and Palria, S. 2014. Site suitability analysis of water harvesting structures using remote sensing and GIS-A case study of Pisangan watershed, Ajmer district, Rajasthan. *The International Archives of Photogrammetry, Remote Sensing and Spatial Information Sciences*, 40(8), 1471.

Quinn, P., Beven, K., Chevallier, P., and Planchon, O., 1991. The prediction of hillslope flow paths for distributed hydrological modelling using digital terrain models. *Hydrological Processes*, 5, 59–79.

Rana, V. K., and Suryanarayana, T. M. V. 2019a. Evaluation of SAR speckle filter technique for inundation mapping. *Remote Sensing Applications: Society and Environment*, 16, 100271.

Rana, V. K., and Suryanarayana, T. M. V. 2019b. Visual and statistical comparison of ASTER, SRTM, and Cartosat digital elevation models for watershed. *Journal of Geovisualization and Spatial Analysis*, 3(2), 12.

Rennó, C.D., Nobre, A.D., Cuartas, L.A., Soares, J.V., Hodnett, M.G., Tomasella, J., and Waterloo, M.J., 2008. HAND, a new terrain descriptor using SRTM-DEM: Mapping terra-firme rainforest environments in Amazonia. *Remote Sensing of Environment*, 112 (9), 3469–3481.

Rizeei, H. M., Pradhan, B., and Saharkhiz, M. A. 2018. Surface runoff prediction regarding LULC and climate dynamics using coupled LTM, optimized ARIMA, and GIS-based SCS-CN models in tropical region. *Arabian Journal of Geosciences*, 11(3), 53.

Ross, C. W., Prihodko, L., Anchang, J., Kumar, S., Ji, W., and Hanan, N. P. 2018. HYSOGs250m, global gridded hydrologic soil groups for curve-number-based runoff modeling. *Scientific Data*, 5, 180091.

Rozos, D., Bathrellos, G. D., and Skillodimou, H. D. 2011. Comparison of the implementation of rock engineering system and analytic hierarchy process methods, upon landslide susceptibility mapping, using GIS: A case study from the Eastern Achaia County of Peloponnesus, Greece. *Environmental Earth Sciences*, 63(1), 49–63.

Saaty, T. L. 1977. A scaling method for priorities in hierarchical structures. *Journal of Mathematical Psychology*, 15(3), 234–281.

Saaty, T. L. 1987. Rank generation, preservation, and reversal in the analytic hierarchy decision process. *Decision Sciences*, 18(2), 157–177.

Saaty, T. L. 1990. How to make a decision: The analytic hierarchy process. *European Journal of Operational Research*, 48(1), 9–26.

Saaty, T. L. 2008. Decision making with the analytic hierarchy process. *International Journal of Services Sciences*, 1(1), 83–98.

Singh, J. P., Singh, D., and Litoria, P. K. 2009. Selection of suitable sites for water harvesting structures in Soankhad watershed, Punjab using remote sensing and geographical information system (RSandGIS) approach—A case study. *Journal of the Indian Society of Remote Sensing*, 37(1), 21–35.

Singh, L. K., Jha, M. K., and Chowdary, V. M. 2017. Multi-criteria analysis and GIS modeling for identifying prospective water harvesting and artificial recharge sites for sustainable water supply. *Journal of Cleaner Production*, 142, 1436–1456.

Sørensen, R., and Seibert, J. 2007. Effects of DEM resolution on the calculation of topographical indices: TWI and its components. *Journal of Hydrology*, 347(1–2), 79–89.

Strahler, A. N. 1957. Quantitative analysis of watershed geomorphology. *EOS, Transactions American Geophysical Union*, 38(6), 913–920.

Tagil, S., and Jenness, J., 2008. GIS-based automated landform classification and topographic, landcover and geologic attributes of landforms around the Yazoren Polje, Turkey. *Journal of Applied Sciences*, 8, 910–921.

Tripathi, S. S. 2018. Impact of land use land cover change on runoff using RS and GIS and curve number. *International Journal of Basic and Applied Agricultural Research*, 16(3), 224–227.

UNESCO-WWAP. 2009. Climate Change and Water, an Overview from the World Water Development Report 3: Water in a Changing World. A World Water Assessment Programme Special Report. UN World Water Assessment Programme, France

Weerasinghe, H., Schneider, U. A., and Loew, A. 2011. Water harvest-and storage-location assessment model using GIS and remote sensing. *Hydrology and Earth System Sciences Discussions*, 8(2), 3353–3381.

Weiss, A.D., 2001. Topographic position and landforms analysis. Poster presentation, ESRI Users Conference, San Diego, CA.

Wilson, J. P., and Gallant, J. C. 2000. Digital terrain analysis. *Terrain Analysis: Principles and Applications*, 6(12), 1–27.

Wu, R. S., Molina, G. L. L., and Hussain, F. 2018. Optimal sites identification for rainwater harvesting in northeastern Guatemala by analytical hierarchy process. *Water Resources Management*, 32(12), 4139–4153.

5

Flood Risk Zone Mapping Using a Rational Model in Highly Weathered Nitisols of Southeastern Nigeria

Daniel Aja, Eyasu Elias, and Ota Henry Obiahu

CONTENTS

Introduction

Flood risk zone mapping is an important first step in the proper management of future flooding events and to develop adequate mitigation

DOI: 10.1201/9781003175018-5

measures (Elmira, 2016). In particular, development of flood vulnerability maps at the local government level can achieve a better result than the conventional national maps because it can identify rural dwellers and small holder farmers that are at risk (Asare-Kyei et al., 2015). Flood vulnerability maps are useful tools in identification of populations and elements at risk and to guide early warning system and preventive measures. They are needed in spatial planning to prevent development in flood-prone areas and for implementation of a flood insurance scheme (De Moel et al., 2009).

West Africa has witnessed frequent floods due to high variability in rainfall patterns, geographic location and general low elevations. In the last three decades, the sub-region has witnessed a dramatic increase in flood events, with severe impacts on livelihoods, food security and damaging properties worth millions of dollars (Armah et al., 2010). In 2007, for example, a series of anomalously high rainfall events caused severe floods which affected more than 1.5 million inhabitants in West Africa. This has resulted in the destruction of farm lands, destruction of infrastructure, outbreak of disease epidemics and the loss of human lives (Braman et al., 2010). In 2012, flooding along the river Niger, the principal river in West Africa, resulted in the death of 81 and 137 people in Niger and Nigeria, respectively, while displacing more than 600,000 people (Integrated Regional Information Network, 2013). The frequency of occurrence of extreme events is expected to increase as a result of projected increase in extreme rainfall that may "have dire consequences for the sub-region's agricultural sector and food security in West Africa" (Intergovernmental Panel on Climate Change, 2014).

Abakaliki Local Government Area (ALGA) is popularly known for rice (*Olivia sativa*) farming in Nigeria because of the availability of large expanse of swampy areas adequate for rice cultivation. The Nigerian Hydrological Services Agency in 2014 listed ALGA of Ebonyi State among the moderate flood risk areas in the country. Every year, farmers lose significant quantities of their farm produce due to inundation of crop fields by seasonal floods. Due to the lack of locally relevant and functional flood hazard maps for mitigation and adequate response to flood hazards in this area, farmers do not have required knowledge of the extent of flood coverage in that part of the country.

Despite the major impact of floods on the livelihoods of the people living in the low-lying regions of West Africa, few attempts have been made to delineate the boundaries of flood intensity to indicate areas that are vulnerable to flooding (Asare-Kyei et al., 2015). The limited research conducted on flood mapping in Nigeria has used remote sensing data aided by Geographic Information Systems (GIS). However, they lack certain basic principles in hydrological modeling and prediction which can be added into flood simulation and mapping in the country for better outcomes (Komolafe et al., 2015).

Here we report on a recent study conducted to explore the application of GIS and some hydrologic models in flood extent mapping, especially for data scarce environments at the community level in ALGA in Southeastern Nigeria. The overall objective of the study was to develop detailed flood hazard maps at a fine spatial resolution with aim of providing information for early warning, risk preparedness and to put in place adequate response mechanisms.

Material and Method

Description of the Study Area

The study was conducted in ALGA of Ebonyi State, Southeastern Nigeria (Figure 5.1). The geographical coordinate lies within 06°04′0″N Latitude and 08°65′0″ E Longitude. ALGA occupies the eastern axis of Ebonyi State, covering a land area of about 584 km^2. The area is characterized by high relative humidity of about 71%–75% and surface temperature of 26°C–31°C with mean temperature of 30.4°C. There is a bimodal rainfall pattern from April to July and September to November with a short spell in August (Figure 5.2) and a long-term average rainfall of 1,296 mm. Hydrologically, the area is located within the derived savannah zone of South-East Nigeria, lying within the plains of Ebonyi River, Iyiudele and Iyiokwu Rivers that are also tributaries of Cross River (Figure 5.1). The inhabitants are predominantly agrarians raising livestock and crops at both subsistence and export levels. Major crops for national and international markets are rice, cassava and yam (Ogbodo, 2013).

Data Sources and Methods

Data Sources

We made use of digital elevation model from ASTER which is a joint product of the Japanese Ministry of Economy, Trade and Industry and the United States National Aeronautics and Space Administration. The data have a vertical accuracy of 17 m at 95% confidence level, and a horizontal resolution on the order of 75 m. The land cover data were obtained from Landsat 8 imagery which was downloaded from the USGS website.

We used soil maps from the Harmonized World Soil Database version 1.2 produced in 2012 by the International Institute for Applied System Analysis

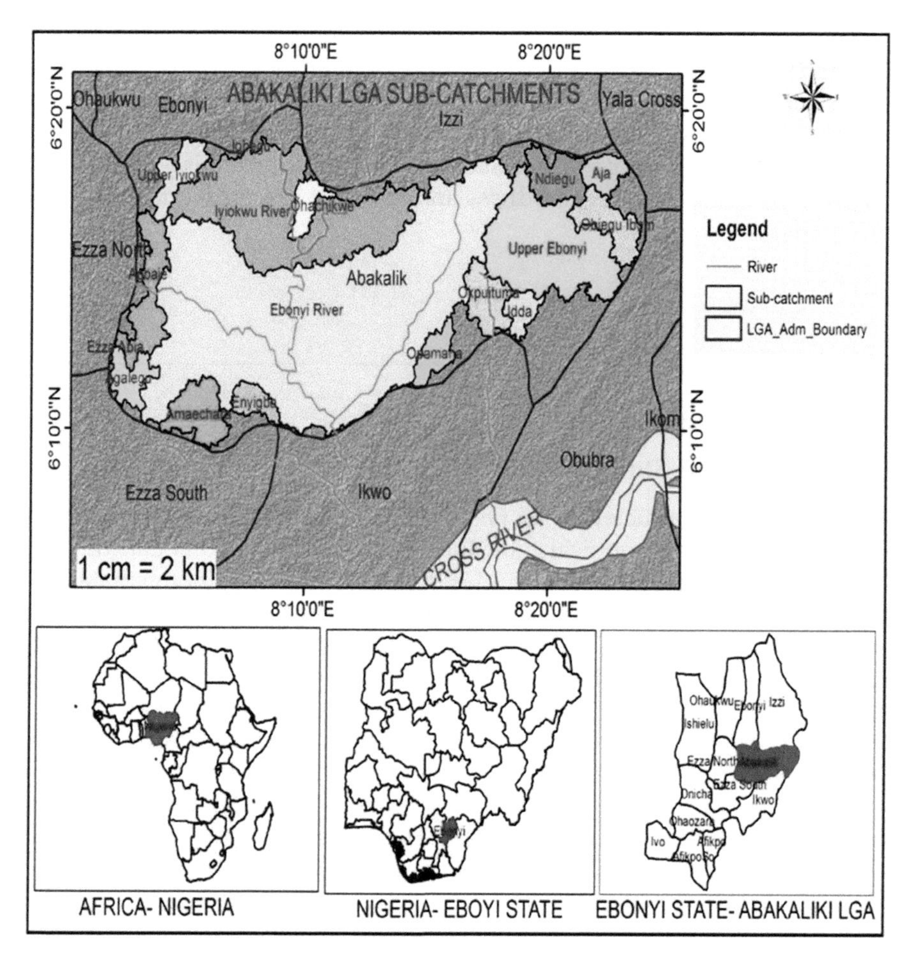

FIGURE 5.1
Map of the study area.

for soil type and texture classification. The soil map has 1 km resolution. We obtained rainfall data from the Nigerian Meteorological Agency, Abakaliki area synoptic station. Topographic maps of Ebonyi State covering the study area and the shape files of the administrative boundary and settlements for the study area were obtained from the Ministry of Lands and Survey Abakaliki.

Run-Off Estimation Models

The methodological approach that was employed is diagrammatically summarized in Figure 5.3 as described by Asare-Kyei et al. (2015). Hydrological

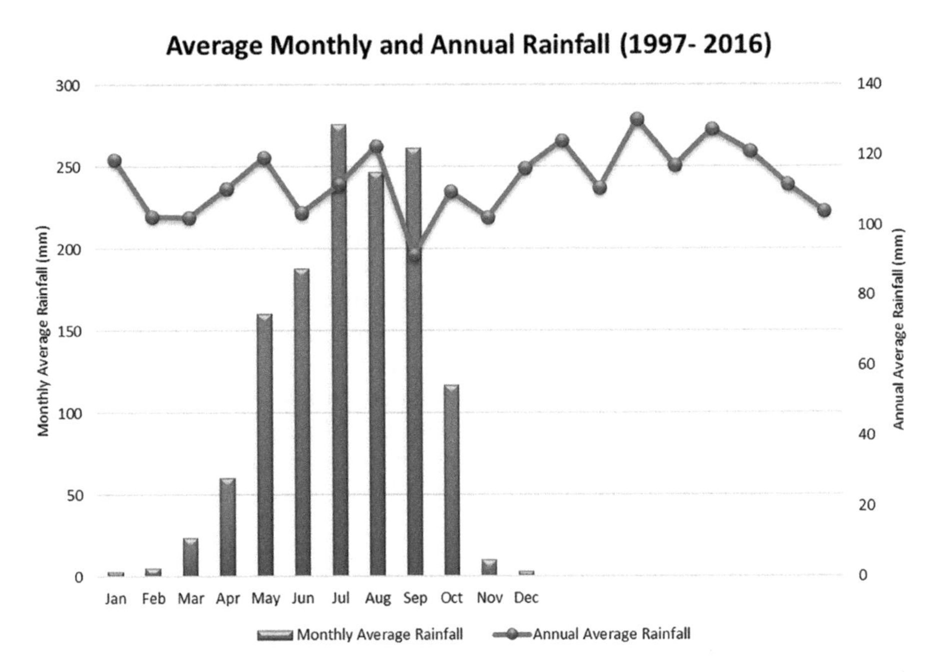

FIGURE 5.2
Seasonal distribution and long-term annual average rainfall in ALGA.

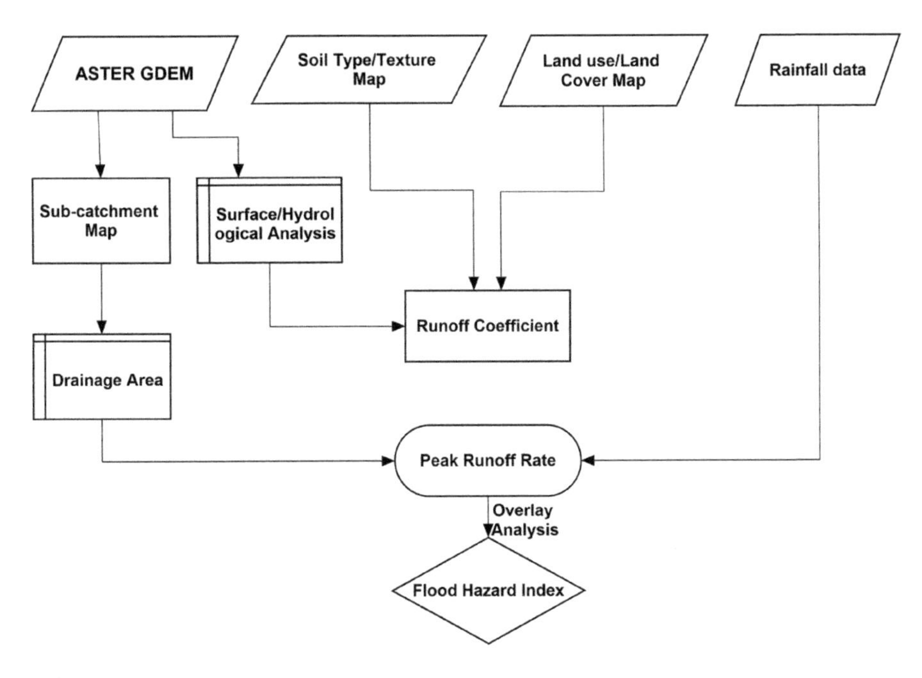

FIGURE 5.3
Modified modeling flow diagram for relational-rule-based flood assessment.

modeling (flood risk zone mapping) was achieved by using combined application of the rational hydrological model and arithmetic overlay operation. First, the study area was delineated into sub-catchments using ArcGIS10.3 software. Secondly, a modified version of the rational model was used to estimate the run-off of the respective sub-catchments based on rainfall intensity, and a run-off coefficient. Finally, the arithmetic overlay operation was applied in a GIS environment to integrate the output of the hydrological model with other flood causal factors such as elevation to determine a flood intensity map for the sub-catchments. Flood-prone zones were eventually defined through a reclassification of the flood intensity map to derive the Flood-Prone Index (FPI) which determines the flood-prone zones of the area. This approach involves retrieving data values from all flood covariates and then calculating peak run-off rates using the rational model. The covariates for flood are land use/land cover (LULC), soil type and soil texture, slope, elevation, rainfall and drainage area (Morjani et al., 2014).

Determination of Peak Run-Off Using the Rational Model

The rational model belongs to the group of lumped hydrological models which treats the unit of analysis as a single unit whose hydrological parameters (e.g., rainfall) are considered as average values. The model is given by the following equation:

$$Q_p = 0.0028 \times C \times I \times A \tag{5.1}$$

where Q_p = Peak run-off rate (m^3/s), C = run-off coefficient, I = rainfall intensity (mm/h), and A = drainage area (ha). A constant (0.0028) is required to convert the original units in North American system (where the model was first developed) to an international system such as cubic meters per second (m/s). The model operates on a number of assumptions including: (1) the entire unit of analysis is considered as a single unit; (2) rainfall is uniformly distributed over the drainage area; (3) estimated peak run-off has the same chances of reoccurrence (return period) as the used rainfall intensity (I); and (4) the run-off coefficient (C) is constant during the rain storm.

The strength of this model lies in its simplicity for application and its suitability for a homogeneous area. As a result, this model has a wide application in the calculation/estimation of peak run-off rate for the design of different drainage structures and flood hazard map production (Nyarko, 2002). The model converts rainfall in the catchment into run-off by calculating the product of the rainfall intensity in the catchment and its area, reduced by a run-off coefficient (*C*, with a value between 0 and 1) which depends on the soil type, land cover and slope in the study catchment. The run-off coefficient provides an estimation of how much rainfall is lost through infiltration, interception and evapotranspiration. This means that the run-off coefficient of a catchment can be seen as the fraction of rainfall that actually becomes run-off. Therefore, accurate estimation of the run-off coefficient is vital to the successful implementation of this method.

Sub-Catchments Delineation

Digital Elevation Model was used for sub-catchment delineation and slope analysis. The study area was delineated into 17 sub-catchments by clicking on the spatial analyst tools in ArcGIS environment after all the sinks had been filled to make it more perfect. The filled elevation data layer was maintained and used later for the integration of peak run-off and elevation to determine run-off concentration at different elevations. The Hydrology tool was expanded to perform various hydrological analyses such as flow direction, flow accumulation, stream order, stream to feature and subsequently sub-catchment determination. The sub-catchments which were generated in a raster format were immediately converted to polygon by clicking on the conversion tool under spatial analyst tools. The conversion of the raster format into polygon was necessary in order to calculate the areas of the sub-catchments and also to build the attribute table in the ArcGIS environment.

Peak Run-Off Map Development

Within each sub-catchment, more than one LULC types and slope exist. To find a run-off coefficient that will represent a given sub-catchment, average values were taken based on the different LULC types. The DEM was also converted to percent slope in ArcGIS and was reclassified into three classes; slope less than 2%; slope between 2% and 6%; and slope greater 6%. Based on tables from Bengtson (2020), which specifies a run-off coefficient for a particular LULC type and slope, the average values of run-off coefficient for each sub-catchment were computed based on the number of LULC that occur in each sub-catchment. Knowing the run-off coefficients (*C*), rainfall intensity (*I*) and areas (*A*) of each sub-catchment within the study area, the discharges (*Qp*) for each likely to cause flooding was obtained.

Flood Covariates and Acquisition Methods

LULC Analysis

LULC maps of the catchment were generated by classifying moderate spatial resolution (30 m) multitemporal Landsat images which were processed prior to analysis. The LULC data were generated for three periods, namely, 1986, 1996 and 2016 exploring changes in the LULC type overtime.

The different bands of Landsat imagery were combined in the ArcGIS environment to form composites, and the composites were further processed into raster mosaics prior to analysis. Supervised classification was conducted on the Landsat imagery to reveal four broad LULC classes after training samples and signatures were created (using the training sample manager), saved and imported into the ArcGIS environment. These land use classes identified were (1) agricultural land; (2) forestland; (3) bare land; and (4) settlements (i.e., built-up areas). Training and validation data for these classes were obtained from field campaigns conducted between December 2017 and March 2018. Training and validation samples for the classification were generated by overlaying the training and validation data (polygons) on the satellite image and extracting the corresponding values.

Soil Type and Texture

The Harmonized World Soil Database which was used for soil classification is an image file linked to a comprehensive attribute database where information on soil mapping units, soil texture for top and sub-soils and several other soil properties are stored (Food and Agricultural Organization, 2009). Based on this information, the extracted soil map of the area was reclassified into the four main soil hydrological groups (A–D) defined by the United States Natural Resource Conservation Service (USDA, 2009).

Integration of GIS Model

The GIS model (GISM) as presented in Figure 5.3 was adopted and modified for analysis (Asare-Kyei et al., 2015). The model uses four main stages for flood risk zoning including (1) the generation of the different maps of the study area using satellite data, elevation map and field survey; (2) the inclusion of these data into the GISM and building of attribute tables; (3) the use of arithmetic overlay operation to combine the hydrological model with the

GIS model; and (4) the creation of flood vulnerability map for the area under study.

Finally, the elevation layer and the peak run-off layer were combined using arithmetic the overlay method in ArcGIS to generate the flood hazard intensity map at different elevations. The model combines DEM and discharge maps within the GIS environment to determine flood risk areas. The arithmetic overlay method involves two main stages:

Determination of run-off concentrations (Figure 5.7a) within various segments over the landscape.

$$X + Y = Z_{ct} \tag{5.2}$$

Estimation of values that can be used to infer potential areas likely to be in flood with any storm event (Figure 5.7b).

$$X + Y / X = Z_{FRA} \tag{5.3}$$

where X (m) is the digital elevation model; Y (m^3/s) represents total discharge; Z_{ct} (m^3/s/m) is the run-off concentration at various elevations; and Z_{FRA} is the value for flood risk areas.

To develop maps for easy understanding, a reclassification was done to redefine five flood hazard intensity categories, viz. very high, high, moderate, very low and low risk zones. The natural breaks reclassification method in ESRI's ArcGIS was used for this purpose (Kazakis et al., 2015; Xiao et al., 2017).

Results

LULC Changes from 1986 to 2016

The LULC changes for the study area between 1986 and 2016 are presented in Figure 5.4. The results show that there have been changes in the various LULC (forest, agricultural lands, bare lands and settlements) from 1986 to 2016. However, only settlements (built-up areas) show significant change from what it used to be over the years (1986–2016). This indicates that there is correlation between LULC and flooding in the area.

In 1986, forestland covered 4,107 ha (7.7%) of the catchment, agricultural land accounted for 38,919 ha (72.6%), bare land accounted for 7,860 ha (14.7%) of the catchment, while 2,730 ha (5.1%) were covered by settlements (built-up areas). In 1996, 3,530 ha (6.5%) of the study area was forest, 41,130 ha (76.7%) of the study area was agricultural land, 5,190 ha (9.7%) of the study area was bare land, while 3,770 ha (7%) of the study area was covered by settlements.

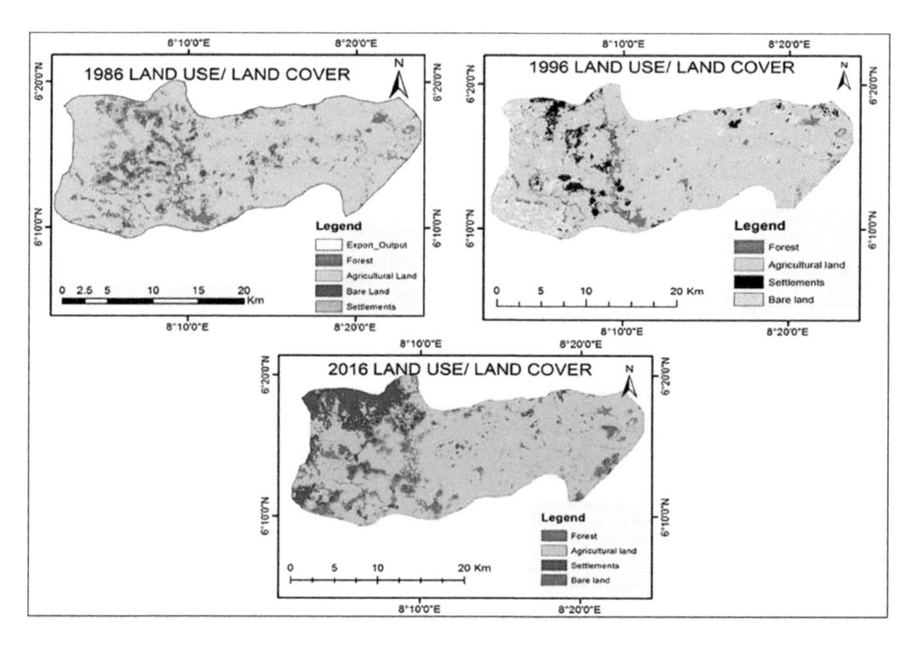

FIGURE 5.4
Maps of LULC classification (1986–2016) for study area.

By 2016, the following observations were made: forest covered 3,720 ha (6.9%), agricultural land covered 38,090 ha (71%), bare land covered 5,300 ha (9.9%), while settlements covered 6,510 ha (12.1%) of the study area.

Soil Textural Class and Elevation of the Sub-Catchments

The result of the soil classification revealed that the study area is predominantly Nitisols (NT) (Figure 5.5) representing the hydrological soil group "C" which is characterized as shown in Table 5.1 below. High elevation values are concentrated in the upper Ebonyi River (60 masl); upper Iyiokwu River and Ezza Abia sub-catchments while lower Ebonyi River, Iyiokwu River and Obiagu Ibom records very low elevation. The lowest elevation (15 masl) was observed in the southernmost part of Ebonyi River.

Peak Run-Off Analysis

The map of the peak run-off rates (m^3/s) shows the distribution of run-off within the sub-catchments in the area studied (Figure 5.6 and Table 5.2). The

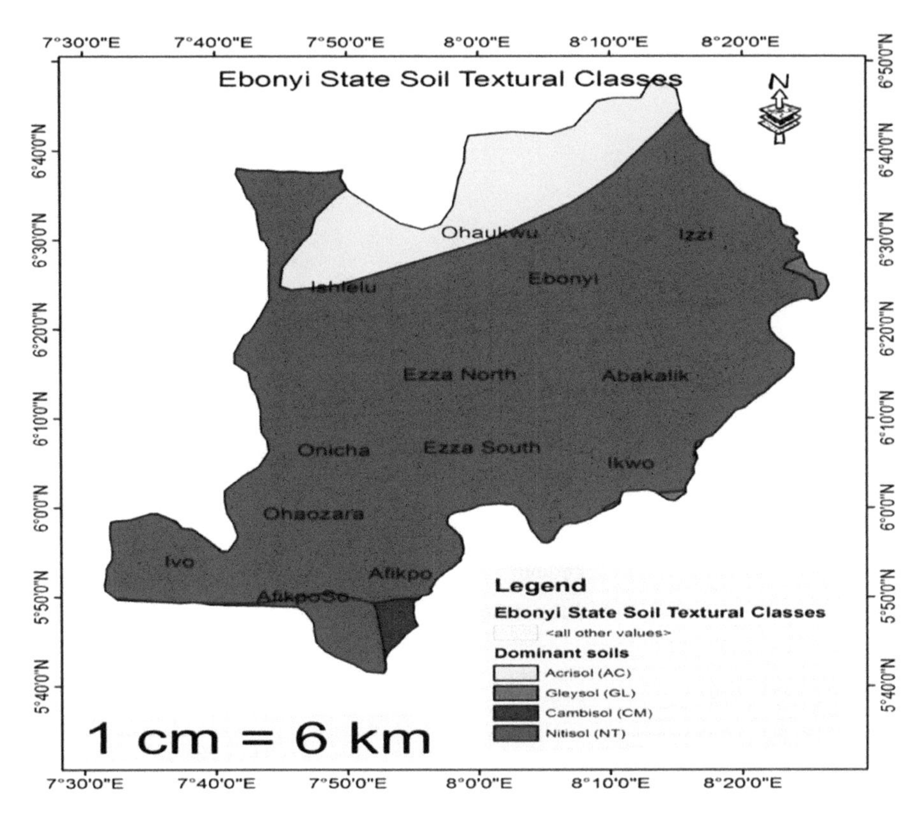

FIGURE 5.5
Map showing soil types in Ebonyi State.

TABLE 5.1
Hydrological Soil Groups

Soli Groups	Infiltration Rate (in/h)	Description	Relative Run-off Potential
A	>30	Sand, Loamy Sand	Low
B	0.15–30	Sandy loam, Loam	Moderate
C	0.05–0.15	Silt Loam, Sandy Clay Loam	High
D	0.0–0.05	Clay loam, Silt Clay loam, Sandy Clay & Clay	Very high

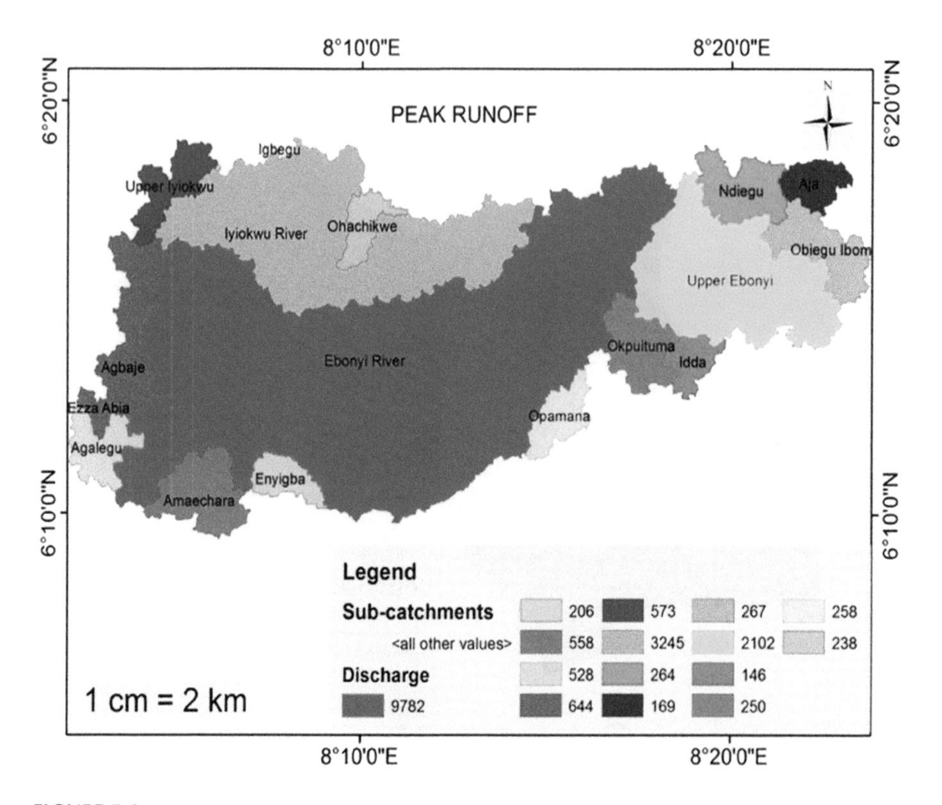

FIGURE 5.6
Map showing peak run-off discharges of sub-catchment.

Ebonyi River sub-catchment generates the highest amount of run-off in excess of 9,782 m^3/s, while the Igbegu sub-catchment generated the lowest (0.03 m^3/s).

Flood Hazard Intensity Map

This map was produced by overlaying the peak run-off layer with the elevation layer through arithmetic overlay method as discussed (Figure 5.7a and b).

A reclassification was done on the flood vulnerable areas map to produce five classes which represent the Hazard Index. The index ranges from 1 (very low flood hazard intensity) in some parts of upper Ebonyi River sub-catchment to 5 (very high flood hazard intensity) in the lower part of Ebonyi River sub-catchment. The final flood hazard map is represented in a graduated color (Figure 5.8). The map shows that about 33% of the catchment falls within very high flood hazard areas that cover sub-catchment such as Ebonyi River, Iyiokwu River and Igbegu. On the other hand, the very low

TABLE 5.2

Sub-Catchment Discharges Based on August 2016 Rainfall

Sub-Catchment	Area (km²)	Run-Off Coefficient (C)	Rainfall Intensity (mm/h)	Discharge (m³/s)
Igbegu	0.07	0.42	414.2	0.03
Ajaa	582	0.25	414.2	169
Ndiegu	1085	0.21	414.2	264
Upper Iyiokwu	883	0.56	414.2	573
Ohachikwe	539	0.38	414.2	238
Obiegu Ibom	1099	0.21	414.2	267
Iyiokwu River	7365	0.38	414.2	3245
Upper Ebonyi	4770	0.38	414.2	2102
Okpuituma	1025	0.21	414.2	250
Idda	503	0.25	414.2	146
Ezza Abia	0.10	0.52	414.2	0.06
Agbaje	1632	0.34	414.2	644
Opamana	654	0.34	414.2	258
Agalegu	747	0.61	414.2	528
Enyigba	538	0.33	414.2	206
Ebonyi River	22195	0.38	414.2	9782
Amachara	1267	0.38	414.2	558

flood hazard areas account for 44% of the study area covering sub-catchments such as upper Iyiokwu, upper Ebonyi, Ndiegu, Okpuituma and Ezza Abia. The very high flood hazard intensity zone is concentrated in Ebonyi River sub-catchment that is characterized by the highest run-offs and the lowest elevation of 9,782 m³/s and 15 m, respectively, resulting in the greater percentage of the sub-catchment to fall within the very high flood hazard zone (Figure 5.8).

Discussion

From the result of LULC change detection between 1986 and 1996, forest and bare land areas decreased by 1.2% and 5%, respectively, while agricultural land and settlement (built-up area) increased by 4.1% and 1.9%, respectively. This shows that forest and bare land areas have been converted to either agricultural land or settlements during this period.

Again, between 1996 and 2016, forest area increased by 0.4% probably due to government intervention via afforestation. During this period, agricultural land decreased by 5.7% while settlements areas and bare land increased by 5.1% and 0.2%, respectively. This increase in areas covered by

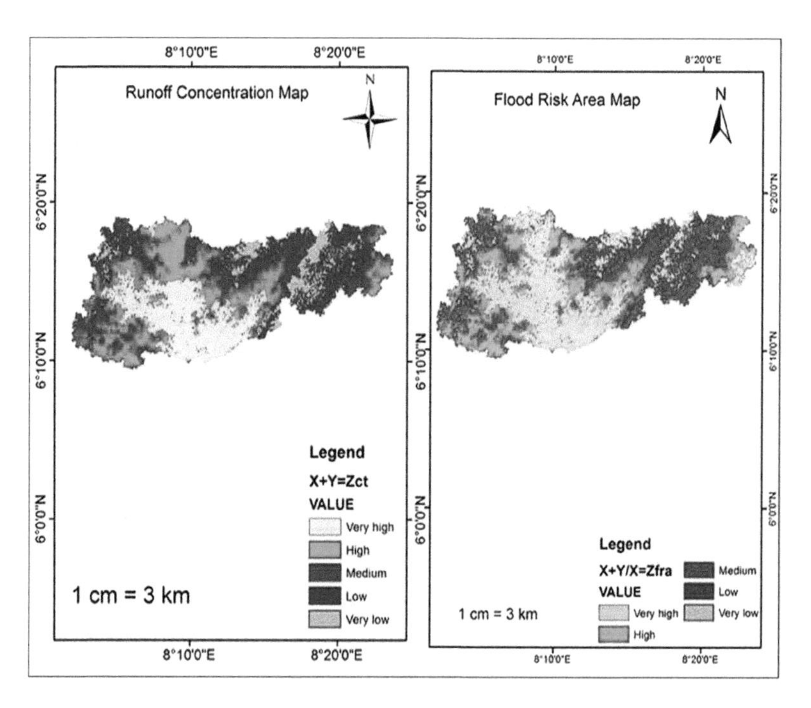

FIGURE 5.7
Maps of run-off concentration (a) and flood vulnerable areas (b).

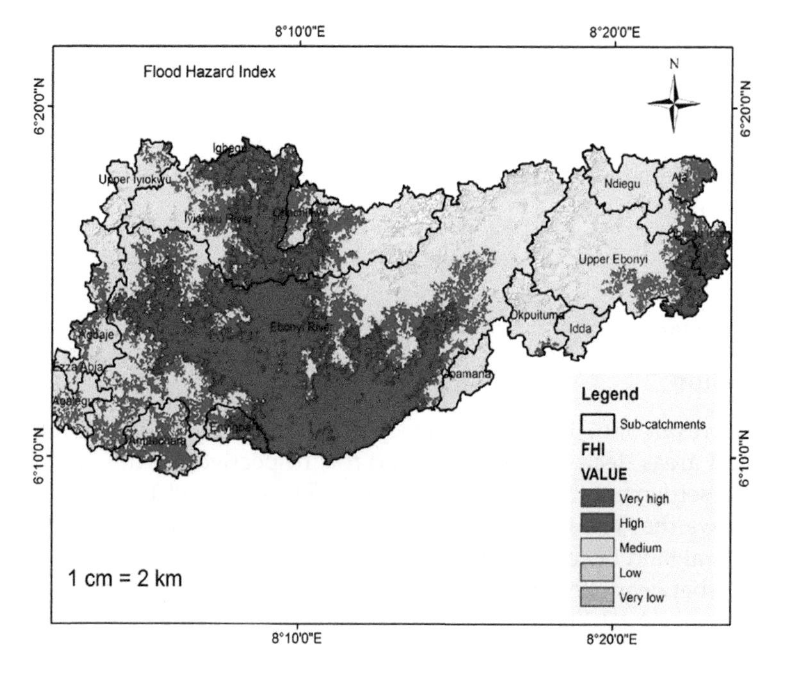

FIGURE 5.8
Flood hazard map.

settlement and bare land between 1996 and 2016 could be attributed to high influx of people to the state capital. It is important to mention that part of the study area became a state capital in 1996 which led to rapid increase in population.

Soil classification based on the soil attributes in the Harmonized World Soil Database shows that the soil properties in the study area influence high run-off generation which can ultimately lead to flooding. Areas with low elevations fall in the category of high flood intensity zone while areas with high elevations fall within low flood intensity zones in the Flood Hazard Index. This indicates that elevation plays a major role in flooding.

High run-off has a positive correlation with increased susceptibility of flood hazards. This is consistent with the key informant interviews with experts which revealed that communities within Ebonyi River and Iyiokwu river sub-catchments experience more frequent flood events and more people suffering from flood impacts when compared to other sub-catchments. As reported in Islam and Sado (2000), the high flood risk in Ebonyi River and Iyiokwu river sub-catchments was related to hydrological parameters.

Our findings through focus group discussions and key informant interviews with community members also revealed that flooding has been an issue in the catchment. The state government in collaboration with the federal government has channelized the two major Rivers (Iyiokwu and Iyiudele) which are responsible for major floods within Abakaliki metropolis in a bid to control the impact of flooding in the area. The channelization was done between 2013 and 2015 through the ecological fund and it has greatly reduced the frequency and severity of flooding within the metropolis. However, other areas which are not within the metropolis continue to witness different intensities of flooding.

Conclusion

Our study elaborated an approach to synthesize the relevant database in a spatial framework to produce a flood vulnerability map of ALGA through the application of simple hydrologic models and arithmetic overlay operations in the ArcGIS environment. Coupling of these hydrological models with GIS and remote sensing techniques in this study has shown the potential for accurate flood risk zone mapping. With this method, flood risk of various land uses can be determined with a greater accuracy. This could allow for more accurate estimation of most flood risk elements and identification of flood safe areas to prioritize developmental efforts. The study identifies rainfall intensity, LULC changes, soil properties and elevations as major factors that influence flooding hazard.

The flood mapping showed that Ebonyi River sub-catchment has a very high flood extent followed by Iyiokwu River, Iyiudele River and Obiagu Ibom sub-catchments. Therefore, early warning system development and mitigation interventions must be put in place in those areas. Accordingly, policy makers and development planners can make use of this study to develop appropriate early warning system and flood mitigation measures and consequently reduce the effects of flooding on the livelihoods of rural small holder farmers by taking note of the spatial extent of flooding in the area. This study provides important information that can be useful for decision-makers to prioritize developmental efforts at local government levels.

We urge agricultural extension workers in the state to step up their game in educating farmers on the use of early maturing species and the importance of upland rice farming to reduce crop inundations by seasonal flooding. Sustainable flood awareness campaigns/programs are encouraged even in periods without flooding to continuously include the culture of resilience on the communities.

A major limitation of this work, however, is that the hydrological model used does not consider some important factors that determine the magnitude of flood such as antecedent moisture conditions. We recommend an assessment of flood depth in further research on the study area to take the above limitations into account.

Acknowledgment

This paper is part of the author's MSc thesis work which was funded by the European Union through Intra-ACP AFIMEGQ Scholarship. Parts of this contribution made use of materials from a previous publication "Aja, D., Elias, E., and Obiah, O.H., 2020, Flood risk zone mapping using rational model in a highly weathered Nitisols of Abakaliki Local Government Area, Southeastern Nigeria. *Geology, Ecology, and Landscapes*, 4(2), 131–139" with permission from publisher Taylor & Francis Group.

References

Armah, F. A., Yawson, D. O., Yengoh, G. T., Odoi, J. O., and Ernest, K. A. 2010. Impact of floods on livelihoods and vulnerability of natural resource dependent communities in northern Ghana. *Water*, 2(2), 120–139.

Asare-Kyei, D., Forkuor, G., and Venus, V. 2015. Modeling flood hazard zones at the sub-district level with the rational model integrated with GIS and remote sensing approaches. *Water*, 7(7), 3531–3564.

Bengtson, H. H. 2020. *Hydraulic Design of Storm Sewers Using Excel. Online Continuing Education for Professional Engineers,* OnLine PDH, Merritt, FL.

Braman, L. M., Pablo, S., and Maarten, K. V. 2010. Climate change adaptation: Integrating climate science into humanitarian work. *International Review of the Red Cross,* 92(879), 693–712.

De Moel, H., Van Alphen, J., and Aerts, J. C. 2009. Flood maps in Europe—Methods, availability and use. *Natural Hazards and Earth System Sciences,* 9, 289–301.

Elmira, B. 2016. Developing a flood risk map a case study of the city of Pori. Finland. Bachelor's Thesis, Novia University of Applied Science in Ekenäs, Finland.

Integrated Regional Information Network. 2013. West Africa Flood Round-Up. http://www.irinnews.org/news/, last checked February 15, 2022.

Intergovernmental Panel on Climate Change. 2014. The IPCCs Fifth Assessment Report (AR5) Synthesis Report, 1–4.

Islam, M. M., and Sado, K. 2000. Development of flood hazard maps of Bangladesh using NOAA-AVHRR images with GIS. *Hydrological Sciences,* 3, 45.

Kazakis, N., Ioannis, K., and Thomas, P. 2015. Assessment of flood hazard areas at a regional scale using an index-based approach and analytical hierarchy process: Application in Rhodope-Evros region, Greece. *Science of the Total Environment,* 538, 555–563.

Komolafe, A. A., Suleiman, A. A. A., and Francis, O. A. 2015. A review of flood risk analysis in Nigeria. *American Journal of Environmental Sciences,* 11(3), 157–166.

Morjani, E., Zine, E., and Abidine, A. 2014. *Methodology Document for the WHO E-Atlas of Disaster Risk. Flood Hazard Modelling,* World Health Organization, Geneva, Switzerland.

Nigeria Hydrological Services Agency [NIHSA]. 2014. 2014 Flood Outlook for Nigeria.

Nyarko, B. K. 2002. Application of a rational model in GIS for flood risk assessment in Accra, Ghana. *Journal of Spatial Hydrology,* 2, 1–14.

Ogbodo, E. N. 2013. Assessment and management strategies for the receding watersheds of Ebonyi State, Southeast Nigeria. *Journal of Environment and Earth Science,* 3, 3.

USDA 2009. Chapter 7: Hydrologic soil groups. In *Part 630 Hydrology National Engineering Handbook,* Natural Resources Conservation Service, Washington, DC.

Xiao, Y., Yi, S., and Tang, Z. 2017. Integrated flood hazard assessment based on spatial ordered weighted averaging method considering spatial heterogeneity of risk preference. *Science of the Total Environment,* 599–600, 1034–1046.

6

Morphometric Indicators-Based Flood Vulnerability Assessment of Upper Satluj Basin, Western Himalayas, India

Amit Jamwal and Vikram Sharma

CONTENTS

DOI: 10.1201/9781003175018-6

Introduction

Globally river basins are known for floods of both natural and human origins due to unsystematic development, changing land cover, and degradation of watersheds. Watershed and basin geomorphology hold great significance for understanding of, and can often foretell, potential actions of hydrological processes. Hence, river morphometric and flood studies include identifying, quantifying, and prioritizing or ranking the process-related vulnerability in systems.

Mitigations through flood management and catchment area treatments are not possible without study of landscape geomorphic and topographic aspects such as slope, relative relief, slope profile, slope aspect, lithology, geology, soil texture, precipitation pattern, land use land cover (LULC) as well as morphometric parameters of basin and region. Traditionally, river basin hydrology has been understood with the study of morphometric parameters and geology of landscapes (Strahler, 1952). The first scientific-based morphometric study was conducted by R.E. Horton in 1945, and the earlier phase was on a small scale and the second phase was marked with improvements of detail on a larger scale (Nag and Chakraborty, 2003; Magesh et al., 2011).

In modern management, flood control process, monitoring, and assessment are possible via applied application of remote sensing and geographic information systems (Minakshi and Goswami, 2014). These techniques and field measurements and modeling have yielded valuable management insights. In recent times morphometric study is no longer complex and/or tedious work as compared to traditional methods. The advent of the digital elevation model (DEM)-based study was very helpful for landscape analyses (Rao, 2010) as now regional land mass morphometric analyses could be done with the help of satellite images (Dabrowski et al., 2008), and all to assist with the assessment of vulnerability.

Flood vulnerability assessments are based on selective parameters including topographic, morphometric, and other physical environmental aspects (Li et al., 2016). Vulnerability basically refers to the potential of the degree of damage that can be done and is often explained through a scale ranging from value

0 (no damage) to 1 (full damage) (United Nations, 1982; Menoni and Pergalani, 1996). The scoring is based on the threat and parameter sensitivities, with threat and vulnerability deciding the risk factor intensity (UNDP, 2004).

Globally a total of 12% of the people died because of floods as compared to other hazardous deaths (Ritchie and Roser, 2021). The Satluj River is known for its devastating floods in years 1973, 1975, 1988, 1991, 1993, 1995, 1997, 2000, 2001, 2003, 2005, 2007, 2009, 2010, 2013, and 2018. Hence, a flood-based assessment of vulnerability was done for this area based on parameters such as basin shape and area, stream frequency, density, drainage texture, absolute relief, relative relief, dissection index, slope, and river channel gradient (Singh, 2004).

Study Area

The study region of the upper Satluj basin extends from 310 30′12″ N to 320 22′16″ N and 770 40′16″ E to 790 13′16″ E. The total area covered by upper basin of Satluj in the Kinnaur district was 6,401 km^2 and relative relief varies from 414 to 6,818 m (Figure 6.1). Tibet exists in the northeast of Kinnaur, and Uttarkashi

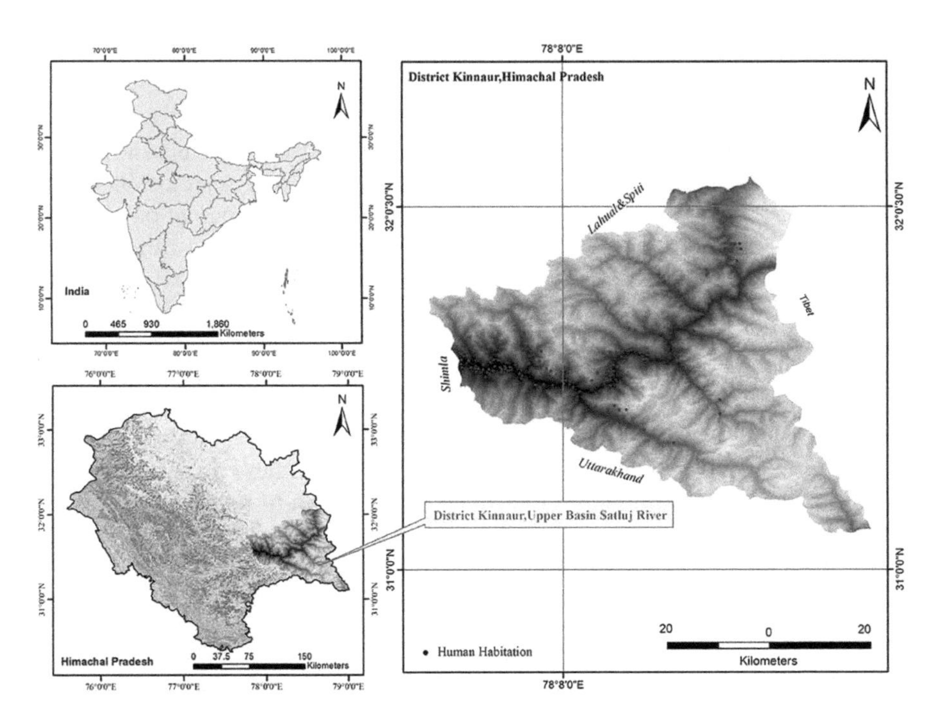

FIGURE 6.1
Study area district Kinnaur, Upper Satluj basin, Himachal Pradesh.

is found south of Kinnaur. This region has a dendritic drainage pattern and a moderate drainage frequency, the density of which is controlled by geological structure. The study region had 8% green or vegetative cover, and a large area of basin was to fall under the barren and snow-covered land cover types. Eighty percent (5,483.92 km^2) of the study region area had high degree of slopes (>300) and 74% (>3,000 m) of the basin area had high altitudes (Jamwal et al., 2019).

The Satluj river basin climate type was classified as sub-humid and arid temperate climates. Sub-humid temperature regions exhibit high density, high texture, high frequency, and high bifurcation ratio, and all receive high incidences of floods. The upper region of the Satluj basin has an arid type of climate and associated drainage frequency, density, and low bifurcation ratio.

This region is known for its social issues related to the physical land degradation and hydropower development (Kuniyal et al., 2017). The upper arid region of the basin received a maximum rainfall of 600–1,400 mm in the months of winter, and the lower sub-humid region received a maximum rainfall in the months of the rainy seasons (July–September). Ten percent of the basin area has suffered huge losses during hazard incidences such as rainfall and flooding (Jamwal et al., 2019) making the issue of vulnerability important for resource management, safety of settlements, and climate.

Methodology

The vulnerability assessment was performed based on selected indicators. The indicators of hydrological and physical aspects were selected based on previous studies (Jamwal et al., 2019). The variety of hydrological parameters included stream length ratio, main stream length, bifurcation ratio, stream frequency, drainage density, infiltration number, drainage texture, length of overland flow, RHO coefficient, compactness constant, channel constant, elongation ratio, circulatory ratio, sinuosity, and ruggedness number. The hypsometric integral and river cross section were derived from the DEM. The physical parameters like river gradients, relative relief, slope, and hypsometric curves were analyzed by using the DEM.

The regional vulnerability was calculated based on these analyzed parameters. Hydrological parameters like river discharge were measured from the pre-installed rain gauges in Tapri and Khab. The river cross section was measured from the DEM and GPS by using the tracking mode. The flood incidences were recorded through field-based observations and secondary data of flood incidences (Kuniyal et al., 2019). Because of complex topography, the Global Positioning System (GPS) points were taken where there was limited possibility of danger and stressing operator safety. The climatic data of the basin was obtained from installed weather stations in Khab, Rekong Peo, and Khab. The historical data were analyzed using the secondary data of Waris India. Then all parameters were analyzed through the hydrological tools of ArcGIS. The values were calculated from pixels in the affected area as measured. The raster layers were overlapped on LULC and the affected area of land cover was measured with the help of ArcGIS 10.3. The LULC data was obtained from the U.S. Geological Survey (USGS) at the resolution of 23m and were processed from the ETM+ 30m coverage. The data were image processed using ERDAS IMAGINE software.

LULC data was classified in eight land use classes via the affected area vector file as overlapped on raster files of LULC. Then affected areas of different land use categories were identified and ranked based on value. Then the ranked value was normalized by using the formula: Normalized value (NV)=Highest ranking value under the classified parameters (HRV)/Total Number of Rank (NR). The main objective of normalized values between 0.1 and 1 was to develop the vulnerability assessment index and final vulnerability map (Jamwal et al., 2019) that was easy to interpret. The analyzed value was converted into point format with their measured and signature values. Then these points were interpolated, one surface was generated, and the surface was categorized as to its value using the weighted overlay analysis. The final vulnerability map was based on a map algebra raster data set with an interpolated surface and classified into low, medium, and high vulnerability classes (Figure 6.2). In the future, the flow diagram of proposed management

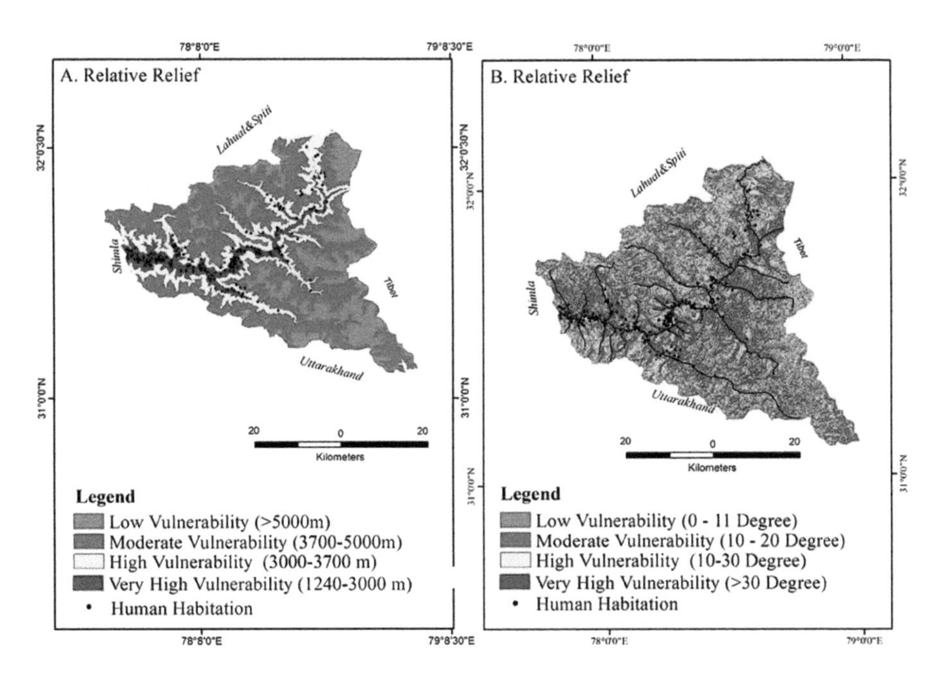

FIGURE 6.2
Relative relief and slope-based vulnerability; (a) relative relief emphasizing within drainage vulnerabilities and (b) relative relief emphasizing outside the drainage vulnerabilities.

information system would be designed based on vulnerability assessments and field surveys.

Hydrological Parameters Analysis

Parameters were collected for study of morphometrics such as stream order (Nu), stream length (Lu), mean stream length (Lsm), bifurcation ratio, stream length ratio (Lms), drainage density (Dd), drainage frequency (Fs), infiltration number (If), drainage texture (*T*), length of overland compactness constant, constant of channel maintenance (C), circulatory ratio form factor (Rf), and RHO coefficient. The first-order streams had maximum stream segments of 1,948, second-order streams had 1,786 segments, third order had 522, fourth order had 224, fifth order had 167, and sixth order had 648 segments, respectively. However, the sixth-order stream has been considered as the mainstream at 55 km in length. The stream length measured in km demonstrated the first-order stream length had 2,212.25 km with second-order segments of 1,049, third order of 559.25, fourth order of 208.29 km, fifth order of 137 km, and the sixth order had 9.43 km, respectively.

The first- and second-order stream quantities indicated the "youthful" stage of the topography. The stream orders followed the next higher order, which indicates erosion and increasing surface runoff of the basin (Nag and Chakraborty, 2003). The basin area was (A) 6,449.03 km^2 and a perimeter of the basin was 386.14 km. The length of the basin (LB) is always parallel to the main drainage, and it was 84 km. The drainage network showed a dendric pattern and the river channel followed the regional slope. At some point, the stream network developed its shape because of hardrock topography and precipitous slopes (Twidale, 2004).

Results and Discussion

Relative Relief

Relative relief is one of the deciding factors of geo hazards and affects the regional geomorphology of drainage system. The high relative relief gives birth to a high dissection ratio. This indicated the high surface runoff and high kinetic energy during the time of rainfall. High relative relief has high river gradient and encourages high vertical erosion.

In the basin the high relative elevational region was highly affected by flash floods. The impact of flooding was high (5) up to the relative relief of

1,240–2,942 m, and the vulnerability scored (1) was also high. The low vulnerability was recorded at higher altitudes of 5,090–6,755 m.

Surface Slope

The slope of the study region was classified into six classes. The maximum affected area was found under the slope categories of 20°–30°. This was ranked by the highest vulnerability (1) and highest impact (2.1 km^2). The flood impact was low on the high degree slope (>50°) and decreased with higher elevation. The 30°–40° slope type had vulnerability scores of 0.5, 40°–50° had scored of 0.3°, and >50° had scored 0.2. The geomorphology of river valley exhibited deep gorges and canyons which were accompanied by high degree of slope.

Stream Length Ratio (Rl)

The stream length ratio starts to decrease with decreasing stream orders, as it depicts the dendritic drainage pattern of stream. The numbers of streams that were developed on the landscape was indicative of uniformity in its shape and length which made clearer the homogeneity of geological structure (Horton, 1945). The proportion of increase of mean length of stream segments of two successive basin orders is defined as the length ratio and was calculated according to the following equation:

$$RL = Lu / Lu - 1 \quad (6.1)$$

where Lu is the sum of the length of all stream segments of the given order and Nu is the number of stream segments of a given order (Strahler, 1952). The first order had scores of 2.1, second order 1.8, third order 2.6, fourth order 1.5, fifth order 1.4, and sixth order 1. The stream length of 100–2,215 m had a vulnerability of 0.5.

Mean Stream Length (Lsm)

Mean stream length (Lsm=Lu/Nu) was measured through the ratio of stream length of order u (Lu) and the total number of stream segment of order u (Nu). The Horton (1945) mean length of channel segments of an existing order was greater than that of the next lower order but less than that of the next higher order. The mean stream length varied from 1.0 to 2.6 km. The mean stream lengths of the first order were 2.6, the second order was 1.8, the third order was 2.6, the fourth order was 1.5, the fifth order was 1.4, and the sixth order was 1. The high vulnerability (1) was found under the mean stream length of 94–100 m.

Bifurcation Ratio (Rb)

Rb is one of the important morphometric parameters that shapes the impact of floods. Bifurcation is the ratio between the total number of stream segments of river order and the number of segments of the next higher order. Flooding has a positive correlation with high values of the Bifurcation ratio (Schumm, 1956).

The bifurcation ratio or Rb = Nu/Nu+1 is where Nu is the total number of stream segments of order u and Nu+1 stream segment of next higher order. The average bifurcation ratio of the basin was 1.5. The basin did not have homogeneous rock nor mature topography. The highest bifurcation ratio was recorded for second-order streams (3.4) and the lowest bifurcation ratio was found for the fifth-order stream (0.2). The first-order score was 1, the third order was 2.1, the fourth order was 1.4, and the fifth order was 0.2 (Strahler, 1964). The variance was measured in the bifurcation ratio which shows that geological structure exhibited dominant control in the bifurcation ratio. The geological elements of folds and faults also affected the value of the bifurcation ratio. The second-order stream bifurcation had a high value of 3.5. For the medium bifurcation ratio, the lithology and vegetation cover had strong control over the bifurcation. The second-order streams had high bifurcation value (>3.4) which indicated the high vulnerability of 1 (Lattif and Sherief, 2012).

Stream Frequency (Fs)

Stream frequency is known as the number of streams per square km of area. The soil texture, LULC, and rainfall pattern were also deciding factors of stream frequency as were the hard rock type lithology, dense forest regions, and dry regions that had low stream frequency (Reddy et al., 2002; Shaban et al., 2005). Sparse vegetation, fine texture, and sub-humid, sub-tropical regions had high intensity of stream frequency (Horton, 1932).

Stream frequency was high in the places like Urni, Shongtong, Karcham Wagtu, Tapari, and Powari regions because of medium to coarse soil texture, sparse vegetation, and excess rainfall during the summer season. The upper region of Pangi, Kalong, Barang, Spillo, Sumdo, Hkab, and Nako had dry regions, snowfall, and maximum portion of dry land type giving birth to the low stream frequency. The average stream frequency was 0.10 which indicated low vulnerability (Ketord et al., 2013). Stream frequency and drainage density had a positive correlation. Stream frequency indicated that the basin had poor vegetation cover and peak discharges owing to low runoff rate. The stream frequency decreased with increasing higher order, and the first order was 0.30 m, the second order was 0.27, the third order was 0.08, the fourth order was 0.03, the fifth order was 0.02, and the sixth order was 0.1. The stream frequency vulnerability score was 1 with impacts of 3.1 and low stream frequency was high at 0.5.

Drainage Density (Dd)

The drainage density of the region indicates the topographic aspect of the landscape. High drainage density demonstrates the very sensitive topography. The high drainage density of the first and second orders indicated that landscapes had weak rock strata and several rills developed. The high drainage density type is prone to soil erosion and high probability of mass movements (Ozdemir and Bird, 2009). The drainage density found under the first-order streams was 0.34 km, the second order was 0.16 km, the third order was 0.08 km, the fourth order was 0.03 km, the fifth order was 0.02 km, and the sixth order was 0.1 km.

Drainage density was the density of the total area of the watershed (Dd = L/A) (Vijith and Sateesh 2006). The low Dd values (0.82 km) represented permeable subsurface material, good vegetation cover, and low relief which caused low flood volumes (Tucker and Brass, 1998; Pallard et al., 2009). It was clear that the higher drainage density was partially from higher rainfall. Increasing drainage density reduced storage volumes and increased flood peak volumes. Flooding was less likely in basins with a low to moderate drainage density and stream frequency (Carlston, 1963); the basin had a very low drainage density with an average density of 2.1. The maximum precipitation occurred in the form of snowfall and vegetation of the region was partially controlled by precipitation patterns. The coarse soil texture was found below the Reckong-Peo and fine texture soil was found above the Reckong-Peo. The high drainage density (1.1) was found under the high drainage density area >1. The low vulnerability score was (0.3) found under the low drainage density area and the medium vulnerability (0.6) was found under the area drainage density of 0.8.

Infiltration Number (If)

The low infiltration value of 0.67 in the region was indicative of the high surface water flow in a very short duration of time (Faniran, 1968). The upper region of the basin had a dry temperate climate, but the lower region had the sub-humid temperate type of climate which receives maximum rainfall in the rainy season (June–September) and the upper region was receiving maximum rainfall in January and February. The lower area of the basin fell under the humid temperate type of climate. Infiltration number was inversely proportional to the infiltration capacity of the basin. The high infiltration value of 0.6 had a high relative vulnerability (1).

Drainage Texture (T)

An important geomorphic concept is the drainage texture by which we mean the relative spacing of the drainage lines (Smith, 1950). The study region had coarse texture which indicated the dominancy of geology and geological elements. The high drainage texture >5 had high vulnerability (1) of soil erosion and mass movement. The drainage texture had values of 0.4–5. The drainage texture of the region was also affected by the climate (Srinivasa Vittala et al., 2004). In the basin, high drainage patterns were developed in the lower portion of the basin because of high rainfall. The sparse vegetation cover and upper basin had a low drainage texture indicative of extreme dry condition and arid temperate type climate.

Length of Overland Flow (Lg)

It is the length of the water flow that runs over the ground before it gets concentrated into the mainstream (Horton, 1932). The overland flow of water is generally determined by the topography, lithology, climate, vegetation cover, and rainfall pattern of the region. Anthropogenic activities also altered the overland flow of water. The low value of overland flow of 0.17 indicated that the basin had high rugged relief with high water movement potential. Its low value indicated the high runoff in a very short period of extreme rainfall (Singh and Singh, 1997). The values less than 0.17 had a low vulnerability score of 0.5, and the high value (0.88) had high vulnerability with a score of 1.

RHO Coefficient

Horton (1945) explained the RHO coefficient as the ratio (RHO=Rl/Rb) of stream length and bifurcation ratio (Rb). The high values of the RHO coefficient determined storage capacity of the basin with both values having a positive correlation. In the Satluj basin the average RHO values were 0.5 and its vulnerability score was 0.3. The highest value was found under the fifth-order streams of the basin (5.6) which had a high vulnerability (1). The high RHO value indicated that the basin had high water storage, high runoff, and chances of high soil erosion loss, which reflects that the basin had high water recharge and storage capacity (Mesa 2006).

Compactness Constant (Cc)

The compactness constant is basically a ratio between the area of basin and the perimeter of basin. The perimeter is the outer area of the basin (Horton, 1945), with the compactness constant directly related to the basin length. Compactness constant has a positive correlation with basin length where CC=0.2821×P/A 0.5 and P=Perimeter of the basin (km) area. The compactness constant of the basin was 1.5 which indicated that the basin was more elongated and characterized by high discharge of water. Its vulnerability score was 1 and directly proportional to the high erosion rate.

Channel Constant Maintenance (C)

Basically, constancy of the channel maintained is one of the shape parameters and inversely related to the drainage density. The high channel constant (0.8) indicated the high permeability of the soil and high structural disturbances (Schumm, 1956). The high vulnerability score was 1 and the affected area was only 3.4 km^2.

Elongation Ratio (Re)

Re is the ratio between the diameter of a circle of the same area as the drainage basin and the maximum length of the basin. The elongated ratio value varies from 0.1 to 0.33. This value indicated that the basin had sparse vegetation cover and high vulnerability of soil erosion. Its elongated ratio indicated that the basin had high surface runoff during short periods of time. Other factors like that of LULC, slope, and drainage pattern influenced its physical degradation (Schumm, 1956; Manu and Anirudhan, 2008). This type of situation with the basin indicated high vulnerability during the time of peak flows of the river. The measured value of 0.1 had high vulnerability (1) and > 0.33 measured value area had a low vulnerability score (0).

Circularity Ratio (Rc)

Miller (1953) showed the circulatory ratio is the ratio of the basin to the area and the perimeter of the basin. The circuitry ratio was affected by the basin relative relief, dissection index, slope, land use, rainfall, and geological structure (Singh and Singh, 1997). The circulatory ratio of the basin was 0.61. However, the more circularity that a basin exhibited presents a greater risk of flash floods and there will be a greater possibility that the entire area may contribute runoff at the same time. The high risk of erosion and sediment load is also associated with the high relief and steep slope (Reddy et al., 2004; Strahler, 1952). The circulatory value >0.61 was indicative of the high changes in surface runoff during catastrophic rainfall events and indicated high vulnerability (1). The circularity ratio (Rc) of the Satluj basin was highly influenced by the hydrological and physical factors such as stream length, frequency of streams, geological structures, LULC, climate, relief, and slope of the basin.

Sinuosity (Si)

The sinuosity index is Si = (X/Lb), where X = average drainage length, and Lb = basin length (Leopold et al., 1964). The sinuosity was based on the two-point distance measurement. It determined how much stream is straight. High sinuosity indicates the high impacts of floods during incidences (Brice, 1984; Ebisemiju, 1994). Where there is high sinuosity and low depth channel with high river cross section (>100 m) this situation leads to higher damage during floods. The sinuosity index value varied from 0.2 to 5.7, the first order exhibited 5.7, the second order exhibited 2.7, the third order had 1.4, the fourth order had 0.5, the fifth order had 0.3, and the sixth order had 0.2 with the average seniority of the main river as 1.8. The average value indicates a non-meandering stream which has a high vulnerability during extreme rainfall. The river was in its youthful stage and vertical erosion was high, the high sinuosity value (>4.0) had high vulnerability score 1 and the sinuosity value of 0.2 had a low vulnerability (0.3).

Ruggedness Number

Ruggedness number is the product of basin relief and drainage density. The high drainage density and high relative relief indicated high vulnerability. The basin had an average medium drainage density; however, the upper region of the basin had low density, while the lower region had high drainage density because of its drastic change in lithology and precipitation. The Rn of the entire Satluj river basin was 6,785 m, pointing to the very high basin relief and its very high ruggedness index value at 675.15 (Figure 6.2). This indicated the high vulnerability (3.6) and high soil erosion and high runoff with a vulnerability value of 1 (Figure 6.2).

Form Factor

The form factor is defined as the ratio of basin area to the square of the basin length (Horton, 1932). The form factor value always should be less than 0.78. Hence the higher the value of the form factor, the more circular the shape of the basin and vice versa. The form factor of the river basin was 0.18, which denotes a highly elongated form as the smaller the value of the form factor, the more elongated will be the basin (Gregory and Walling, 1973). The low Ff value reveals that it had less side flow for a shorter duration and high flow for a longer duration (Reddy et al., 2004). The high form factor (>0.40) had a high vulnerability score (1).

The Channel Slope Gradient

Channel slope gradient is dependent on the elevation of the basin; if the region has high relative relief, then the basin has a high gradient. The basin elevation varied from 1,244 to 6,755 m, with a high average slope gradient of 11° indicating that the basin was in its youthful stage and vertical erosion is high. Leopold, Wolman, and Miller (1964) stated that the mean channel slope decreased with increasing successive order in

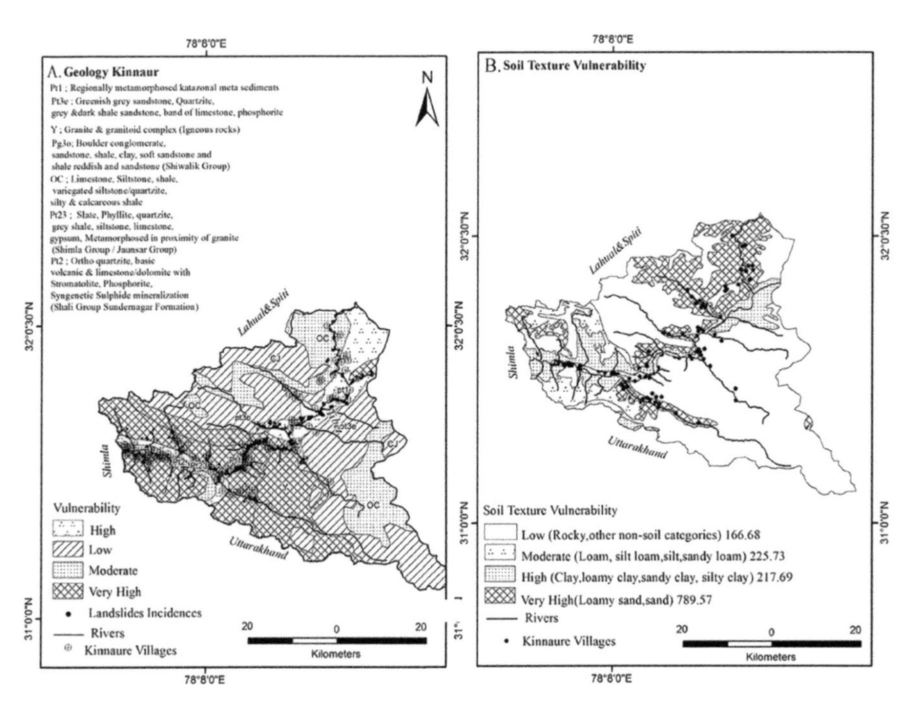

FIGURE 6.3
Vulnerability of soil texture and lithology; (a) Geology Kinnaur and vulnerabilities and (b) soil texture vulnerabilities

geometric series with constant slope ratio. The lowest slope gradient was recorded at 0.20 and the highest slope gradient was at 570. The channel has a high slope gradient which directly relates to potential of vertical erosion. The slope average gradient was high which indicates the potential of high risk during the time of floods, but the slope gradient was near to the cross-section places which had high risk as the settlements were very near to the river bank (Figure 6.3). The channel slope gradient of 0–300 had a high affected area with high vulnerability (1.0). High channel slope gradients are one of the deciding factors for the development of hydroelectric projects.

The Hypsometric Curve (HC) and Hypsometric Integral (HI)

The hypsometric curve was obtained by using percentage height (h/H) and percentage area relationship (a/A) (Luo, 1998). The relief aspect of the basin is related to the study of three-dimensional features of the basin involving area volume and altitude (Pike and Wilson, 1971). The hypsometric curve (HC)

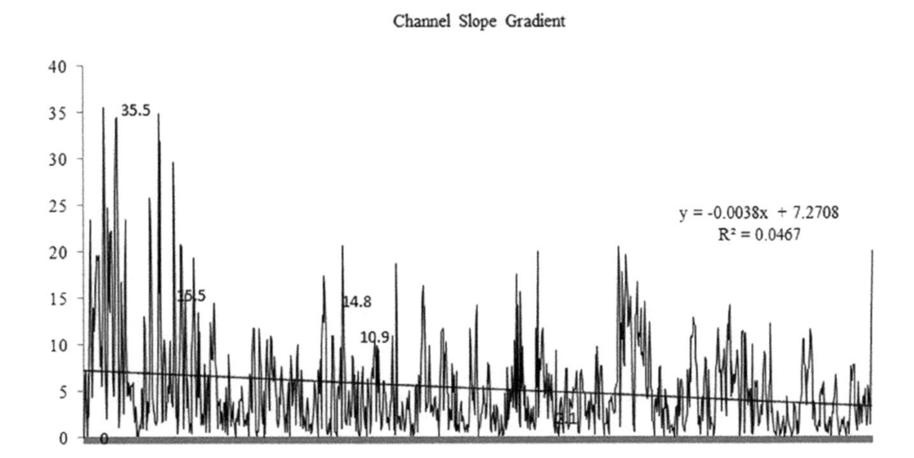

FIGURE 6.4
Channel slope gradient of river upper Satluj basin in district Kinnaur.

and hypsometric integral (HI) were calculated by using a digital elevation model. Approximate values of relative relief were utilized to plot the hypsometric curves for the watershed, from which the HI values were calculated using the elevation-relief ratio method elaborated by Pike and Wilson (1971). The hypsometric calculation involves the measurement and analysis of altitude and basin area to understand the degree of dissection and the stage of the cycle of erosion. Hypsometric Curves are generally used to show the proportion of surface area at various elevations above or below a datum (Smith and Clark, 2005). It is represented by a cumulative percentage and reveals the actual profile of the terrain (Luo, 1998). The maximum area of >80% of the basin falls under the high altitude of the basin of >3,000 m. The 22% hypsometric integral average area indicated the basin is in its "youthful" stage of development (Figure 6.4). The youthful stage basin had a high vulnerability of flash floods and related erosion. Because the basin has a high relative relief, and the youthful stage of the river has a high capacity to carry the sediments during the time of extreme rainfall (Rai et al., 2017a). Generally, the Re values deviated from 0.6 to 1.0 for most of the basins. The value ranges from 0.6 to 1 for regions having a high relief with a circular shape (Magesh and Chandrasekhar, 2012). The hypsometric integral had a high vulnerability score of 1.

River Cross-Section Measurement

River cross-section measurements included the river width measurement, along with the river channel and its bed measurement. The river sites namely Tarna, Nathpa Jhakri, Wagtu, Tapri, Shoultu, Choling, Karcham, Shongtong Powari, Skiba, Morang, Up Mphal, Spillo, Punag, and Khab had high elevated topography where cross-section measurements were taken from the satellite images (Figure 6.5). Locations such as Tarna, Shongtong, Skibba, Morrang, Upmohal, Spillo, Punag, and Khab had extended river cross section with gentle river gradient of <5° (Sui and Koehler, 2001; Figure 6.6). In these places river gradients are very high and cross-sections of these places were narrow (60–70 m). The narrow cross-section was found in the river valleys where geomorphology of the river had long and narrow gauges. These locations had less vulnerability of human and social infrastructure, but physical landscape vulnerability was very high. The flood incidences were seen in all these locations, but the vulnerability of natural and human resources was less (Figure 6.4). Tapri, Nathapa, Jhakri, Karcham, and Powari were the places where the river bed was near the location and often only 10–20 m away. The structural management was not found on every location of the basin. The high cross section of 100–200 m of the river had high vulnerability (1). The cross section with values of 100 m had medium vulnerability (0.5), and the low cross section (<100 m) had low vulnerability score of 0 (Figure 6.5).

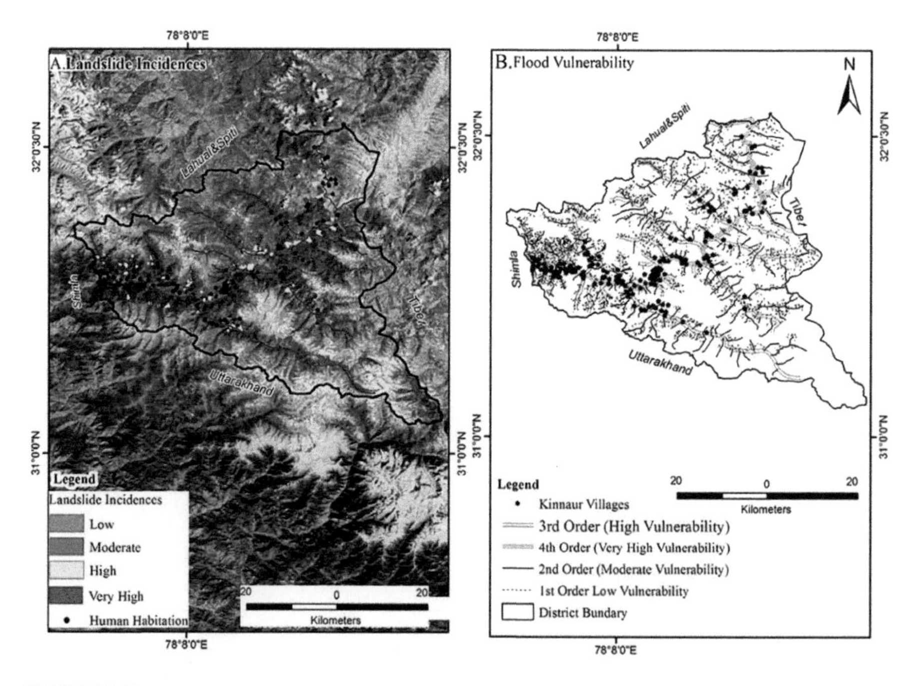

FIGURE 6.5
Landslide incidences and flood-based vulnerability; (a) landslide incidences and their frequencies, and (b) flood vulnerabilities and stream orders.

River Discharge

It was revealed from historical evidence that Rampur, Tapri, Shoultu, Shongtong, and Powari settlements suffered during the time of floods. The river discharge depends on the intensity of rainfall. The determining factors such as soil, slope, relative relief, coarse texture, high degree of slope, and high dissection increase the impact and vulnerability of physical loss. The major flood incidences were recorded in years 1973, 1975, 1988, 1991, 1995, 1997, 2005, and 2014 (Gupta and Sah, 2008a). The average water discharge of the basin was 434 m^3 (Figure 6.6). The maximum water discharge recorded in 2010 was 1,306.42 m^3/s (Figure 6.7). In this basin, the water discharge greater than 800 m^3/s becomes a flood (Gupta and Sah, 2008b). The secondary data revealed that the flow rose above the level of flood vulnerability in 2008, 2011, 2010, and 2014. The vulnerability score of 1 was high at high discharged value (>1,000 cm^3/s). The loss of landscape was high (3.6) with a high value of 1 (Figure 6.7).

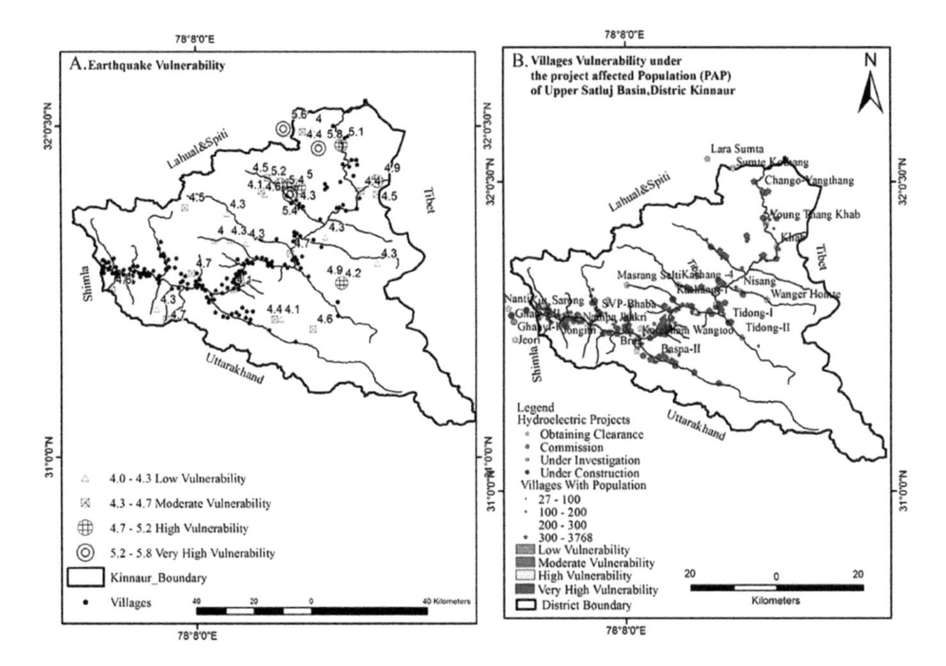

FIGURE 6.6
Earthquake-based vulnerability and project-affected population vulnerability; (a) earthquake vulnerability levels, and (b) villages vulnerabilities under projects.

FIGURE 6.7
(a) Lose unconsolidated soil texture near to Akpa village, (b) large landslide in Bara Khamba, (c) muck dumping at the river Satluj side of Karcham Wangtoo HEP, (d) drilling the hill at 100 mw Tidong Hydro, (e) Urni landslide, (f) large landslide at Rekong Peo, (g) questionnaire survey at village Kwangi, (h) houses affected by the landslide at village Nigulseri, (i) strategic environmental assessment meeting at deputy commissioner office Rekong Peo, Kinnaur, on November 2014, (j) construction site of Tidong project, (k) dumping of muck along the river Sainj by Parbati HEP, and (l) house cracks in Yulla village because of the tunnel construction of Karcham Wangtoo HEP.

Temperature and Rainfall

Temperature and precipitation data of 100 years plus were analyzed from 1900 to 2002. Data was based on different weather stations from Khab to Rampur. The temperature was continuously increasing in the study area of the upper Satluj basin. The maximum temperature was recorded as 14.37°C and the minimum temperature was recorded as 11.58°C. Based on temperature data analysis, the basin temperature has increased at the rate of 0.1099°C/year (Figure 6.8). The flood incidences were evident in the region of 800–900 m with a high vulnerability score of 1.0, while the precipitation region of 500–700 cm has low vulnerability (0.3). The precipitation varied from 700 to 800 cm with a medium vulnerability score of 0.6. Precipitation was not increasing with the increasing temperature. The maximum precipitation was recorded as 141.821 cm and the minimum was recorded as 48.46 cm. However, the mean of precipitation was 80.529 ± 18.30 cm. Precipitation has been decreasing at a rate of 0.012 cm/year. Tropical Rain Mission Measure (TRMM) analyzed data from 1997 to 2016 and the average rainfall was found to be 753.72 mm which was highest in 2010 (1038.8 mm). However, the lowest was recorded in 1999

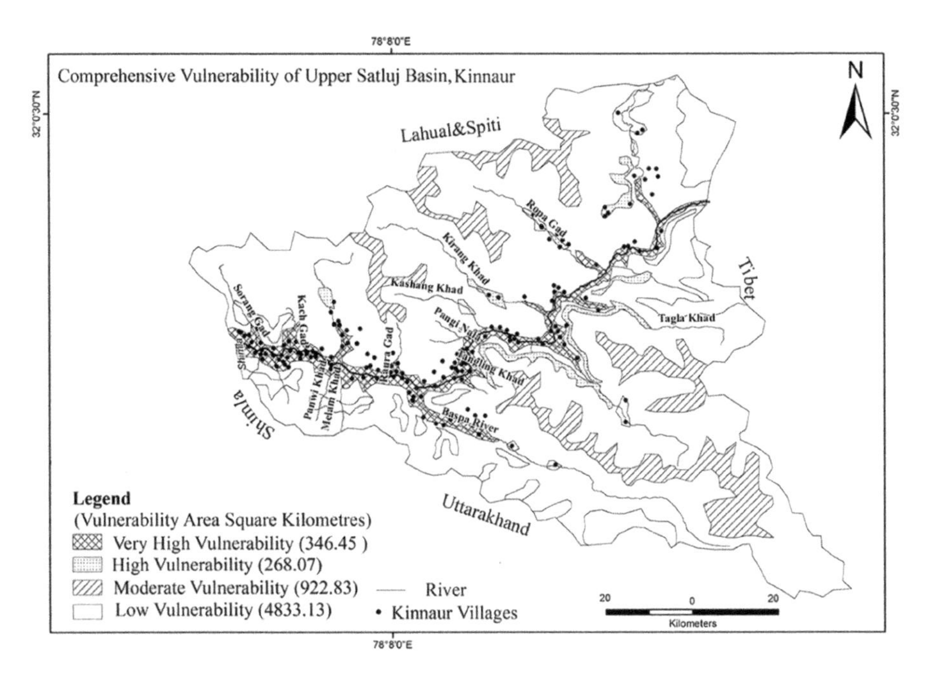

FIGURE 6.8
Comprehensive vulnerability.

(549.6 mm). The average rainfall was 716.25 mm from 1998 to 2008, while the average rainfall recorded was 818.06 mm from 2008 to 2016.

Land Use and Land Cover

LULC helps determine the impact of surface runoff and rainfall. The sparse region has high vulnerability of soil erosion and mass movement. The flood impact was highly controlled in the dense forest region. Tidal mangrove forests protect the river banks from the impact of divesting floods. The Satluj basin in Kinnaur district had very poor vegetative cover at only 8% indicating the high vulnerability of landscape. The land use type's wasteland (1), forest (0.7), and built-up (0.8) were affected with flood incidences with a high vulnerability score. The grassland (0.3) and snow-covered (0.1) had a low affected area with low vulnerability score. The medium vulnerability score was found under the scrubland (0.6) and plantation/orchards (0.4). The maximum area of the study region was to fall under the land use categories of wasteland and snow-covered, with the wasteland (1), scrubland (0.6), and forest land (0.7) having the highest vulnerability score.

Regional Flood Vulnerability

Regional flood vulnerability was analyzed based on the observed values of selective parameters. The scale factor value was fixed from 0 to 1. The value 0 indicates low vulnerability, and when it goes nearer to 1, then the vulnerability starts to increase. The scale was manually fixed from 0.1 to 0.3 as having low vulnerability; the values varying from 0.3 to 0.6 had medium vulnerability and values above 0.6 had high vulnerability. The low vulnerability region 3,936 km^2 (59%) has low value of selected indicators, while this region has very high relative relief of >300 m and was covered with dense forest and snow-covered. Human settlements were not found in this region. The human infrastructure roads and some settlements were found away from the main river and its tributaries. The medium vulnerability regions (23%, 1,516 km^2) had some unbalanced observed values, where high values of physical parameters were encountered such as bifurcation ratio, river gradient, high RHO coefficient, high infiltration, as well as vulnerable land cover and land use. The medium vulnerability region was highly affected with incidences of slope failures and mass movement because of high runoff and high steep slope gradient. The high vulnerability region 16% or 1,112 km^2 was found along the main river valley, and some highlighted spots were severely affected with the floods during the years of 1973, 1975, 1988, 1991, 1995, 1997, 2005, 2014, and 2018 (Figure 6.9).

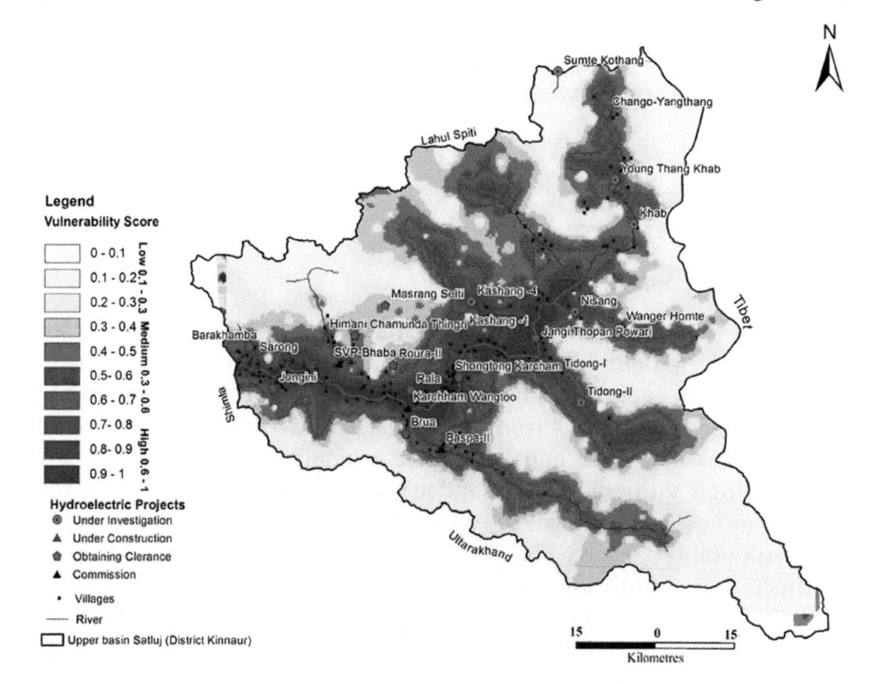

FIGURE 6.9
Vulnerability regions and vulnerability scores in color codes for the upper basin of Satluj. Locations of hydroelectric projects are provided.

Final Outcomes and Suggestive Measures

The vulnerability assessment of basin indicates that the basin has flood vulnerability. The impact of flood is different compared to the plain area flood. The flood impacts were high during the time of excessive and sudden rainfall. However, most of the settlement were away from the river bank, but some settle near the river bank and these settlements were highly affected by floods. This valley had high indication of vulnerability parameters with an average value (0.67) indicative of mass movements, soil erosion, and slope failures. The geological structure revealed that the basin had high presence of geological elements such as mass movement and slope failures as recorded (places like Urni, Shongtong, and Sudharangdhaku) under the thrusts and faults. The analysis of stream orders indicated that stream of third and fourth orders had high vulnerability and high incidences of floods. The places along the river banks like Nathpa Jhakri, Wagtu, Tapri, Karcham, Powari, Telangi, Skibha, and Moorang were highly affected during the time of floods. In developed countries there are warning systems based on the threshold value of water level, but in Satluj valley, there were no warning systems along the river valley. Because flash flood forecasting is very important from a risk management and evacuation point of view, implementation of information management and warning systems, if possible, would help prevent the loss of human life and properties.

The high spatial resolution (0.5 m) satellites datasets can be helpful to understand landslides. Interpretation and geophysical analysis of 5 m resolution DEM can be good inputs for vulnerable maps to generate data for the information management system. The advance metrological stations can be installed in the highly vulnerable areas identified using the vulnerable maps. Thingspeak (external web server) application can be used to storing and retrieve data through the use (Hypertext Transfer Protocol (HTTP)) of the local area network (LAN). Node, MCU will be installed in the flood-prone areas, which will act as the transmitting unit driven by an ultrasonic sensor that detects water levels. When water crosses the threshold value, the messages will be generated from the server. After that, the signal will feed the LAN and then messages will be generated to the local people through the android app on their mobiles (GPRS communications). Thingspeak will generate signals to the voice tower. The vulnerable places can be evacuated during the time of the high alert. However, we cannot reduce the hazards, but we can reduce their impacts through strong management information and warning systems. There is also a possible implementation of wireless sensor networks using flood monitoring systems (based on Zigbee technology).

Landslide monitoring and early warning systems can be developed through use of earth observation data, early warning systems (e.g., Landslide Early Warning Systems or LEWS), as well as rainfall threshold-based modeling and ground-based wireless instrumentation. The active landslides area can be placed on high alert by installing Automatic Rain Gauges, a Wireless Sensor Network, Micro-Electro-Mechanical Sensors (e.g., accelerometer, soil moisture sensor, force sensor, tilt sensor), and accelerometers. These instruments are linked with the servers and continuously transmit a signal to the local network and the main server. Once a threshold value is reached, almost all monitoring networks automatically notify via alerts to operational units, using mainly SMS and/or e-mails services.

An effective management information system can potentially alert people through mobile and warning towers. A better internet network and advanced android applications are implemented in hazard-prone areas to achieve this. The functioning of an effective management information system requires the awareness, participation of local people, and disaster management authorities.

Funding Statement

This study is part of author's Ph.D. (2016–2020) topic, "Hazards and vulnerability assessment of upper Satluj basin for sustainable development, Western Himalayas, India." The authors received research grants (2014–2017) from the G.B. Pant National Institute of Himalayan Environment, Himachal Regional Centre (HRC, Mohal, Himachal Pradesh, India, 175126).

Acknowledgments

We are thankful to the Kumaun Nainital University and Banaras Hindu University for research assistance. The authors are also thankful to the local people of Satluj basin, Kinnaur district during the time of field visits.

Parts of this contribution made use of materials from a previous publication "Jamwal, A., Kanwar, N., and Kuniyal, J.C. 2019. Use of geographic information system for the vulnerability assessment of landscape in upper Satluj basin of district Kinnaur, Himachal Pradesh, India. *Geology, Ecology, and Landscapes*, 4(3), 171–186" with permission by publisher Taylor & Francis Group.

References

Carlston, C.W. 1963. Drainage Density and Stream Flow. U.S. Geological Survey Professional Paper, 422p.

Ebisemiju, F.F. 1994. The sinuosity of alluvial river channels in the seasonally wet tropical environment: case study of River Elm, South Western Nigeria. *Catena* 21:13–25.

Faniran, A. 1968. The index of drainage intensity: a provisional new drainage factor. *The Australian Journal of Science* 31:328–330.

Gregory, K.J., and Walling, D.E. 1973. *Drainage Basin Form and Process: A Geomorphological Approach*. Wiley, New York, pp. 1–456.

Gupta V., and Sah, M.P. 2008a. Spatial variability of mass movements in the Satluj Valley, Himachal Pradesh during 1990–2006. *Journal of Mountain Science* 5(1): 38–51.

Gupta, V., and Sah, M.P. 2008b. Impact of the trans-Himalayan Landslide Lake Outburst Flood (LLOF) in the Satluj Catchment, Himachal Pradesh, India. *Natural Hazards* 45(3): 379–390.

Horton, R.E. 1932. Drainage basin characteristics. *Transactions of the American Geophysical Union* 13:350–361.

Horton, R.E. 1945. Erosional development of streams and their drainage basins: Hydrophysical approach to quantitative morphology. Geological Society of American Bulletin 56: 275–370.

Ketord, R., Tangtham, N., and Udomchoke, V. 2013. Synthesizing drainage morphology of tectonic watershed in upper watershed (Kwan Phayao Wetland Watershed). *Modern Applied Science* 7 (1) 13–37.

Lattif, A.A., and Sherief, Y. 2012. Morphometric analysis and flash floods of WadiSudr and WadiWardan, Gulf of Suez, Egypt: using digital elevation model. *Arabian Journal of Geosciences* 5: 181–195.

Leopold, L.B., Wolman, M.G., Miller, J.P. et al. 1964. *Fluvial Processes in Geomorphology*. Freeman, San Francisco, pp. 1–522.

Li, C., Cheng, X., Li, N., et al. 2016. A framework for flood risk analysis and benefit assessment of flood control measures in urban areas. *International Journal of Environmental Research and Public Health* 13(8): 787.

Luo, W. 1998. Hypsometric analysis with a geographic information system. *Computational Geosciences* 24 (8): 1815–821.

Magesh, N.S., Chandrasekar, N., and Soundranayagam, J.P. 2011. Morphometric evaluation of Papanasam and Manimuthar watersheds, parts of Western Ghats, Tirunelveli district, Tamil Nadu, India: a GIS approach. *Environmental Earth Sciences* 64 (2):373–381.

Manu M.S., and Anirudhan, S. 2008. Drainage characteristics of Achankovil River Basin, Kerala. *Journal of the Geological Society of India* 71:841–850.

Menoni, S., and Pergalani, F. 1996. An attempt to link risk assessment with land use planning: recent experience in Italy. *Disaster Prevention and Management* 5:5–6.

Mesa, M.L. 2006. Morphometric analysis of a subtropical, Andean Basin (Tucuman, Argentina). *Environment Geology* 50:1235–1242.

Minakshi, B., and Goswami, D.C. 2014. Study for restoration using field survey and geo-informatics of the Kolongriver, Assam, India. *Journal of Environmental Research and Development* 8(4): 997–1004.

Nag, S.K., and Chakraborty, S. 2003. Influence of rock types and structures in the development of drainage network in the hard rock area. *Indian Society of Remote Sensing* 31(1): 25–35.

Ozdemir, H., and Bird, D. 2009. Evaluation of morphometric parameters of drainage networks derived from topographic maps and DEM in point of floods. *Environment Geology* 56:1405–1415.

Pallard, B., Castellarin, A., Montanar, A. et al. 2009. A looks at the links between drainage density and flood statistics. *Hydrology and Earth System Sciences* 13:1019–1029.

Pike, R.J., and Wilson, S.E., 1971. Elevation relief ratio, hypsometric integral and geomorphic area altitude analysis. *Geological Society of America Bulletin* 82:1079–1084.

Rao, K.N. 2010. Morphometric analysis of Gostani River Basin in Andhra Pradesh State, Indian using spatial information technology. *International Journal of Geomatics and Geosciences* 2:179–187.

Reddy, O.G.P., Maji, A.K., Gajbhiye, K.S. et al. 2002. GIS for morphometric analysis of drainage basins. *Geological Survey of India* 11:9–14.

Ritchie, H. and Roser, M. 2021. Natural disasters: our world in data. https://ourworldindata.org/natural-disasters, last checked December 24, 2021.

Schumm, S.A. 1956. Evolution of drainage systems and slopes in Badlands at Perth Amboy, New Jersey. *Geological Society American Bulletin* 67:597–646.

Shaban, A., Khawlie, M., Abdallah, C., et al. 2005. Hydrological and watershed characteristics of the El-Kabir River, North Lebanon. *Lakes and Reservoirs Research and Management Journal* 10:93–101.

Singh P., and Kumar, N. 1997. Effect of orography on precipitation in the Western Himalayan region. *Journal of Hydrology* 199:183–206.

Singh, S. 2004. *Geomorphology*. 5th ed., Kalyan Publication, Allahabad. pp. 267–296.

Smith, K.G. 1950. Standard for the grading texture of erosional topography. *American Journal of Science* 248:655–668.

Smith, M.J., and Clark, C.D. 2005. Method for the visualization of digital elevation model for landform mapping. *Earth Surface Processes and Landform* 30 (7):885–900.

Srinivasa Vittala, S., Govindaiah, S., and Honne Gowda, H. 2004. Morphometric analysis of sub watersheds in the Pavagada area of Tumkur District, South India using remote sensing and GIS technique. *Journal of Indian Society Remote Sensing* 32:351–362.

Strahler, A.N. 1952. Quantitative geomorphology of erosional landscape, International Geological Congress. *Algiers* 3(3):342–354.

Strahler, A.N. 1964. Quantitative geomorphology of the drainage basin and channel networks. In Chow, V.T. (ed.) *Handbook of Applied Hydrology*. McGraw-Hill, New York, pp. 4–74.

Sui J., and Koehler, G. 2001. Rain-on-snow induced flood events in Southern Germany. *Journal of Hydrology* 252(14): 205–220.

Tucker, G.E., and Brass, R.L. 1998. Hill slope processes, drainage density and landscape morphology. *Water Resources Research* 34:2751–2764.

UNDP 2004. *Reducing Disaster Risk, a Challenge for Development.* Bureau for Crisis Prevention and Recovery, New York, pp. 1–50.

United Nations. 1982. Proceeding of the seminars on flood vulnerability analysis and on the principles of floodplain management for flood loss prevention, Bangkok, pp. 47–67.

Vijith H., and Sateesh R. 2006. GIS based morphometric analysis of two major upland sub-watersheds of Meenachil river in Kerala. *Journal of the Indian Society of Remote Sensing* 34 (2):181–185.

7

Remote Sensing Measures of Sandbars along the Shoreline of Sonadia Island, Bangladesh, 1972–2006

Md. Enamul Hoque, Sayedur Rahman Chowdhury, Md. Zahedur Rahman Chowdhury, Mohammad Muslem Uddin, Mohammed Shahidul Alam, and Shyamal Karmakar

CONTENTS

Introduction

Coastal areas are places where geomorphological changes happen quickly because of the slow but steady rise in sea level, the intense dynamic processes, such as wave action, that cause erosion and sediment to move (Ritter et al., 1995; Bird, 2011; Kirwan and Megonigal, 2013; Oyedotun et al., 2018; Özpolat and Demir, 2019; Franco-Ochoa et al., 2020; Ghosh et al., 2020; Grases et al., 2020 and Rizzo and Anfuso, 2020); and continuous sediment supply from the upper terrestrial regions (Islam et al., 2002; Hoque et al., 2021). Changes in coastal geomorphology are a global occurrence (Franco-Ochoa et al., 2020). Approximately 70% of the world's coastline

DOI: 10.1201/9781003175018-7

has been eroded during the previous several decades, while less than 10% has been degraded, and the other 20% or so have stayed reasonably stable (Bird, 1969 and 2011; Ghosh et al., 2015). Both erosion and accretion across the coast of Bangladesh are so continuous (Islam, 1991; Islam et al., 2002; 2010; 2016; 2020; Sarwar, 2005; Sarwar and Woodroffe, 2013; Ghosh et al., 2015; Salauddin et al., 2018; Mullick et al., 2019; Crawford et al., 2021) that morphological changes are taking place from time immemorial either as erosion or deposition, like in the case of any other depositional and erosional coasts in the world (Curray and Moore, 1971; Biswash, 1978). We have rather scanty records covering only a few decades about those morphological changes around the coast of Bangladesh (Islam, 1991; Islam et al., 2002; Sarwar, 2005; Islam et al., 2010; Sarwar and Woodroffe, 2013; Ghosh et al., 2015; Islam et al., 2016; Salauddin et al., 2018; Mullick et al., 2019; Islam et al., 2020; Crawford et al., 2021).

Bangladesh is a deltaic plain formed by three great rivers, the Ganges, the Brahmaputra, and the Meghna, which are together referred to as the GBM (Datta and Subramanian, 1997; Ahmad, 2003; Mukherjee et al., 2009 and Islam et al., 2010). The GBM and other major rivers originating from the Himalayas situated in the north flow across the Bengal delta to end up into the Bay of Bengal (Mukherjee et al., 2009; and Khan et al., 2019). Denudation of the Himalayas is extremely active, occurring at a rate of approximately 70 cm per thousand years and resulting in the formation of the world's largest delta (Curray and Moore, 1971; Biswash, 1978; Allison and Kepple, 2001; Mikhailov and Dotsenko, 2007), which transports more than 2.4 billion tons of sediments per year into the Bay of Bengal by GBM (Coleman, 1968, 1969; Curray and Moore, 1971; Biswash, 1978; Viles and Spencer, 1995; Allison and Kepple, 2001). These sediments interact with the Bay of Bengal's dynamic process, leading to coastal morphological changes (Ramage, 1971; Fein and Stephens, 1987, Ali, 1989; Akter et al., 2016). The endless redistribution and cycling of sediments and the seasonal additions because of monsoon floods produce a complex erosion and accretion pattern, i.e., coastal morphological changes (Barua, 1991). A large amount of sediments is also thought to be carried by undercurrents into the deeper Bay of Bengal and the Indian Ocean (Allison and Kepple, 2001; Mikhailov and Dotsenko, 2007). The dynamic processes in the northern Bay of Bengal, especially in the coast of Bangladesh (Murty and Henry, 1983, Khandker, 1997), are significantly influenced by the bottom topography of the Bay which results in frequent morphological changes in the adjacent coast and islands (Siddiqui, 1988; Barua, 1991; Viles and Spencer, 1995 and Hoque et al., 2013).

Moheshkhali Island is located on the north-western side of Cox's Bazar District of Bangladesh (Islam and Hoque, 1999). Sonadia Island is an extended part of Moheshkhali Island comprising active tidal flats (Khan et al., 2005) primarily formed by sandy sediment deposition (Figure 7.1).

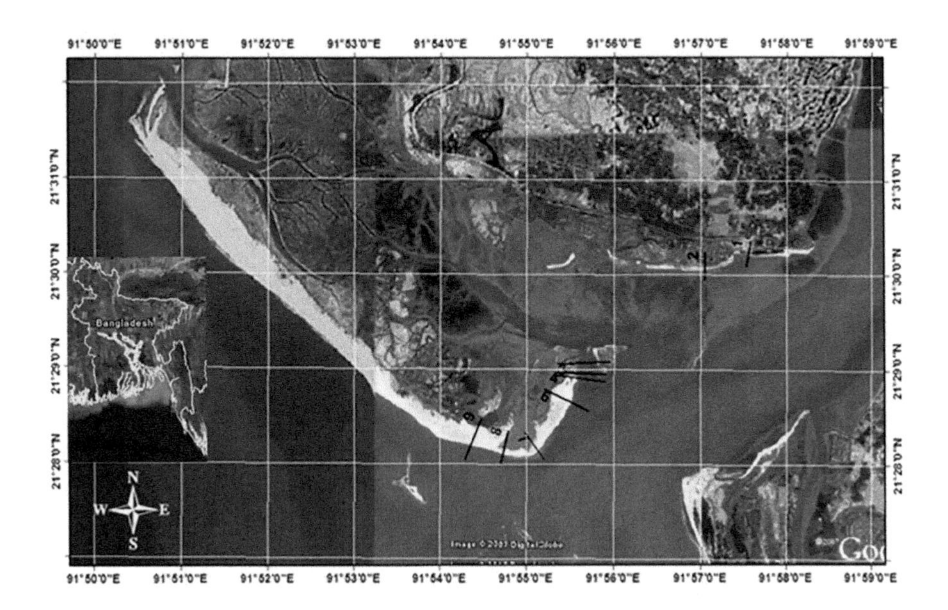

FIGURE 7.1
Map of the Sonadia Island, Moheshkhali, Bangladesh, showing the shoreline, sandbar, and Transects (3–9).

Some khals (creeks) separate it from Moheshkhali Island (Islam et al., 2012; Salam and Hasan, 2019). Though the island is a little far away from the active delta formation region, it still receives substantial sediment (Hoque et al., 2013). It undergoes coastal processes that facilitate reshaping the island's coastline, particularly the south and southeastern part, including Sonadia Island (Hoque et al., 2013). Visible morphological changes are evident within the southeastern shoreline of Sonadia Island, which therefore sees sandbar formation (King and Williams, 1949; Hoque et al., 2013; and Nelson et al., 2013). A study on the formation and morphodynamical changes of sandbars in this geographical location seems essential as it can help understand dynamic coastal processes, including wave and tide and sediment supply (McNinch, 2004 and Pape et al., 2010; and Badrun, 2017). Knowledge of sandbars dynamics is also crucial for sustainable coastal zone management, lacking which may bring failure to any development (Tătui and Constantin, 2020). Widely accepted research has been accomplished on sandbar identification, morphology and dynamics and quantification of sandbars crest positions (Lippmann and Holman, 1990; Ruessink et al., 2002; Lafon et al., 2004; Almar et al., 2010; Aleman et al., 2015, 2017; Tătui et al., 2016; Walstra et al., 2016; and Yuhi et al., 2016; Tătui and Constantin, 2020). Time series of satellite images can reveal the nature and magnitude of changes taking place within this sandbar shoreline.

Thus, the attempt was undertaken based on the analysis of multi-temporal satellite images of the study area acquired between 1972 and 2006 supported by field observations. This study's main aim was to see the morphological changes of a developing sandbar along the southeastern part of Sonadia Island and determine the direction and magnitude of morphological changes over time.

What Is a Sandbar?

A buried or partially exposed ridge of sand or coarse silt produced by waves offshore from a beach is known as a sandbar. Active waves hitting a sandy shore and bottom with whirling turbulence can excavate deep depression by washing away huge unconsolidated sediments part of which later settles on the shore and remaining major parts deposited off the newly formed depression. In addition, suspended sediment coming through the backwash and rip currents and along with the deep-water currents also continue to supply sediments to the newly developed sandbar to make it gradually higher and larger. Such a fully developed sandbar may finally be exposed to form beaches that become connected to the mainland or barrier bars separated from the mainland by small lagoons (Jackson, 2005; Britannica, 2020).

Changes in Sandbar Morphology over Time

Morphological change is a continuous process. Here changes occur in the form of offshore and onshore sandbar migration (Hoefel and Elgar, 2003). Interannual net offshore sandbar migration has been seen in many places across the globe, where bars grow in the inner nearshore, move seaward over the surf zone, and finally degrade offshore in cyclical patterns (Walstra et al., 2012; Cohn et al., 2014).

Wave conditions greatly contribute as incoming large waves break on the sandbar and drive a strong offshore-directed current (undertow) during storms. Gradients in the undertow strength cause erosion onshore, deposition offshore of the sandbar crest, and therefore offshore bar movement (Svendsen, 1984; Faria et al., 2000; Rafati et al., 2021). This causes feedback between waves, currents, and morphological change that pushes the bar offshore until circumstances differ (Thornton et al., 1996; Gallagher et al., 1998; Masselink et al., 2016). In an altered situation, lesser waves do not break on the sandbar but are propelled to the coastline and decelerated by their gentle back faces with the weak undertow. The cross-shore gradients in acceleration skewness cause offshore erosion and onshore bar migration. The sandbar creeps onshore, causing feedback between waves, currents, and morphological change that forces the bar onshore until circumstances shift (Elgar et al., 2001; Hoefel and Elgar, 2003; Hsu et al., 2006).

Sandbar Change Studies Using Remote Sensing Analysis

Materials and Methods

Nine (09) selected transects roughly perpendicular to the shoreline (Figure 7.1) covering the whole sandbar were selected using a Global Positioning System (GPS) survey. The transects were chosen randomly at an interval between 1 and 5 km except Transects 1 and 2. During field observation and visits, sediment samples, beach profiling across the transects and geographic position fixes were collected and analyzed. The sediment sorting results and beach profiles were used to study the relative wave-energy condition presented and discussed earlier in Hoque et al. (2013). Transects 1 and 2 were not situated in the sandbar or Sonadia Island but used for ground-truthing and sediment studies earlier. They are on the Moheshkhali Island, close but separated from Sonadia Island by a tidal creek. Hence, the remaining seven transects (3–9) representing the sandbar were studied to detect the shoreline change.

Satellite Images with Source

Three Landsat satellite images, one ASTER image, and a browse image of QuickBird Satellite over a period between 1972 and 2006 were used in this study to analyze changes in the Sonadia's shoreline. The acquisition time and year of each satellite image are shown in Table 7.1.

TABLE 7.1

Satellite Images Used in the Study of Morphological Changes of Sonadia Island

Year	Satellite & Sensor	Acquisition Date	Resolution
1972	Landsat MSS	03 November	79 m
1989	Landsat TM	22 February	30 m
1999	Landsat ETM	19 December	30 m
2004	Terra ASTER	16 February	15 m
2006	QuickBird	Undated	2.8 m

Landsat satellite images have been acquired from the image archive of the Global Land Cover Facility of the University of Maryland, USA. The ASTER image was ordered from the EROS (Earth Resources Observation System) Data Center's Distributed Active Archive Center (EDC-DAAC), a joint NASA-USGS data archive; and the QuickBird browse image was captured from the Goggle Earth application interface.

Image Analysis

Landsat and ASTER images came with a gross globally accurate georeferencing; however, it was not adequate for a local application like the present study. Therefore, they have been freshly georeferenced using ENVI Remote Sensing Analysis software, taking the Landsat ETM image as the base (reference) image. The QuickBird browse image was also georeferenced in the same process. For the image-to-image georeferencing, distinct landscape features like road junctions and bends, stable ground features, small human-made ponds and ditches, etc. have been used to collect Ground Control Points (GCPs) in the reference image, and they were matched on the other images for corresponding GCPs. Based on a carefully taken set of 15 GCPs, the images were rectified using Polynomial Image transformations. The Root Mean Square Errors (RMSE) of the image corrections process for each image is shown in Table 7.2.

After successful georeferencing, the desired part of the image was cropped (subseted) for subsequent analysis (Figure 7.2).

In Figure 7.2, five cropped images (a–e) of the study area between 1972 and 2006 were overlapped, and the outline of the island was drawn on a single page from each image (f); (a) graphic showing shallows and beach areas from Landsat MSS 1972; (b) graphic showing shallows and beach areas (white to pink) and higher areas in green from Landsat TM 1989; (c) graphic showing shallows and beach areas (white to pink) and higher areas in green from Landsat ETM 1999; (d) graphic showing shallows and beach areas (white to pink) and higher areas in green from Terra ASTER 2004; (e) True color image

TABLE 7.2

Image-to-Image Geo-Rectification Methods and Accuracies for Various Satellite Images

Image	Number of GCPs	Method of Image Transformation	RMSE (pixel)	RMSE (m)
Landsat MSS, 1972	15	Polynomial	0.723	21.69
Landsat TM, 1989	15	Polynomial	0.441	13.23
Landsat ETM, 1999	Reference Image			
Terra ASTER, 2004	15	Polynomial	0.426	12.78
QuickBird, 2006	15	Polynomial	0.461	13.83

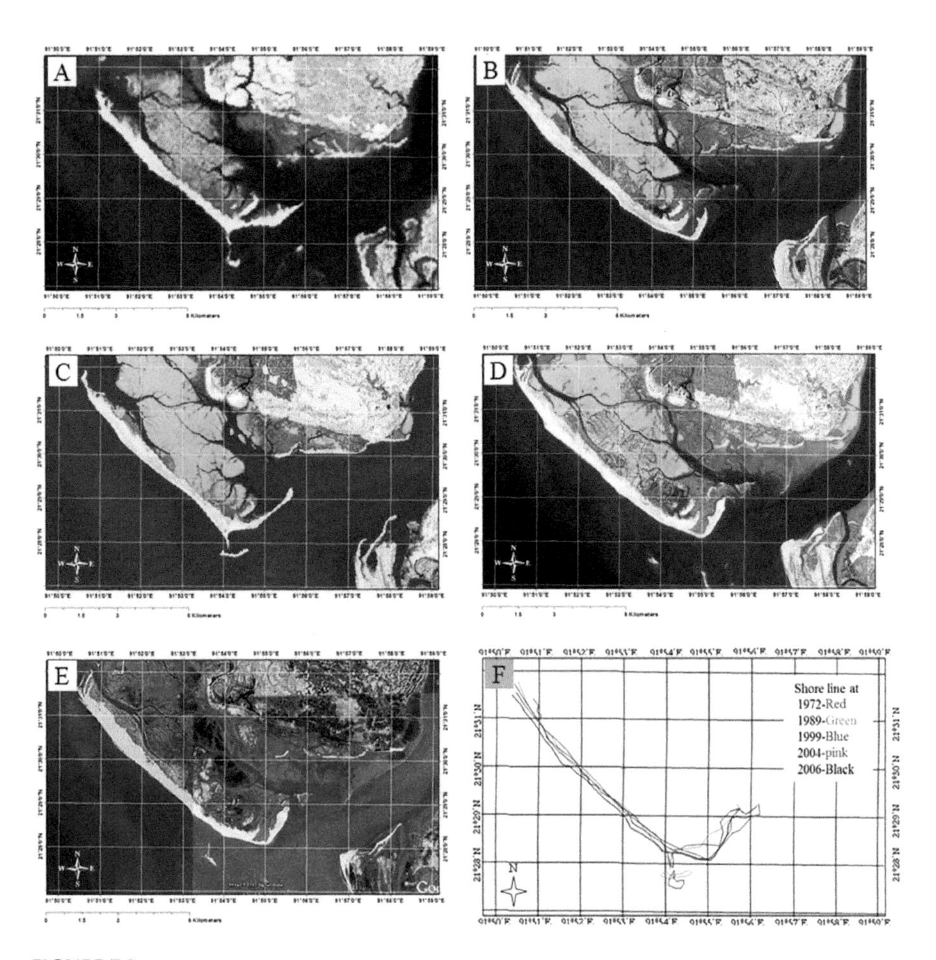

FIGURE 7.2
Georeferenced satellite images of Sonadia Island acquired from different sources and at various years (a–e) and overlapped outline of the southern shore of Sonadia Island of those images (a). Landsat MSS 1972 (b). Landsat TM 1989 (c).

of the barrier island study area from Quick Bird 2006; (f) Shoreline shapes and extent for five years of change.

GIS Analysis of Morphological Change

Five cropped images (Figure 7.2a–e) of the study area between 1972 and 2006 were overlapped, and the outline of the island was drawn on a single page from each image (Figure 7.2f). Morphological changes were determined from those outlines. A line near to perpendicular to most of the shorelines was drawn first, and the distance of each shoreline at each Transect from that perpendicular line was measured manually. The differences between two consecutive images' shorelines were calculated by subtracting the values of the previous image from its successive next image. This difference signifies morphological changes. The values could be found as positive or negative.

Results

This study revealed that the outer shoreline changes (Figure 7.2f) its position from time to time, indicating regular morphological changes between 1972 and 2006. Recurrent morphological changes in the size and shape of the sandbar were observed throughout the period. Here morphological changes included erosion, accretion along with the inward and outward movement of the developing sandbar of the southwestern shoreline of Sonadia Island. The study conjointly found that if erosion was observed along the island's sandbar once, immediately afterwards deposition of sediment accumulates along the sandbar. This phenomenon was common throughout the study area, hence in the entire shoreline of the island. Besides, the sandbar also had a net upward shifting toward the north over time. From keen observation, it was also clear that the sandbar has almost a stable shore in the south and southwestern side of the sandbar, while the eastern part of the sandbar is protruding toward the northeast direction rapidly.

The study area changes in the form of horizontal shifting of the shoreline positions across the transects in various years interval are given in Table 7.3 and Figure 7.3. The average annual rate of displacement of shoreline was 3.96, 3.94, 7.79, and 7.07 m during the 1972–1989, 1989–1999, 1999–2004, and 2004–2006 periods. The highest annual displacement of the shoreline was 22.00 m, which was found at Transect 3 during 2004–2006. The lowest annual displacement of the shoreline was 0.50 m, located at Transects 7 between 1999 and 2004. The changes were rapid between 1999 and 2004 and between 2004 and 2006 than between 1972 and 1989 or between 1989 and 1999. The magnitudes of changes were high at Transects 3, 4, and 5, while the changes were subtle at Transects 6, 7, 8, and 9, i.e., in the sandbar's crest region. These spatially variable rates of changes indicate that morphological changes are high in the southeastern part, i.e., the active sandbar region through shifting of sandbar toward the north and net accretion throughout the periods except net erosion around the southwestern shoreline of the island between 1989 and 1999.

Morphological changes including erosion, accretion, and inward and outward movement of the study area in different periods are also shown in Figure 7.3.

TABLE 7.3

Total Shifting of the Sandbar Along the Shoreline of Sonadia Island in Meters during Various Periods Intervals (Negative Values Denote Erosion While Positive Value Denotes Accretion)

	Periods			
Transect	**1972–1989**	**1989–1999**	**1999–2004**	**2004–2006**
3	94	−26	56	−44
4	72	−36	42	−15
5	−105	−25	38	−13
6	−96	6	53	−3
7	−30	93	−2	11
8	12	87	−16	6
9	24	30	−11	7

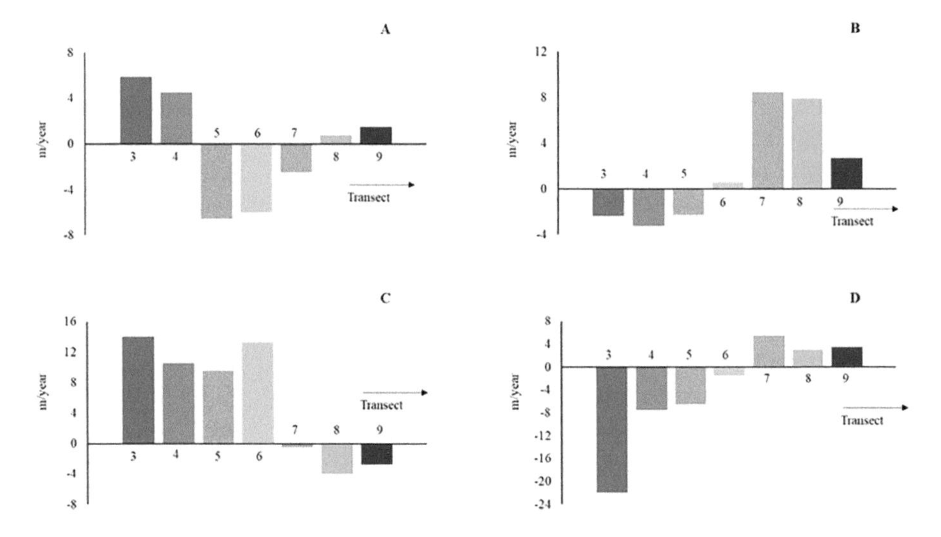

FIGURE 7.3
Average annual rate (meters/year) of changes of the sandbar along the shoreline of Sonadia Island across the transects (3–9) in between various periods. (a) Between 1972 and 1989, (b) between 1989 and 1999, (c) between 1999 and 2004, (d) between 2004 and 2006).

Discussion

Remarkable morphological changes of the sandbar along the shoreline of Sonadia island were observed between 1972 and 2006. Morphological changes include both erosion and accretion of the sandbar and the shoreline of Sonadia Island; along with the sandbar's inward and outward movement. Net accretion on the island and across the coastline during the period was evident. This phenomenon is common in some other coastal regions enriched with heavy sedimentation (Bird, 1969; Lafon et al., 2004; Pape et al., 2010). Once the island along the sandbar undergoes erosion, it accretes in succeeding periods. This pattern of erosion followed by accretion is also seen on other Bangladeshi coastal islands, and is consistent with the results of SPARRSO (1987 and 1993), Pramanik (1983 and 1988), and Pramanik et al. (1981). A few other recent studies suggest such inward and outward movement because of seasonal to annual dynamics of the sandbar morphology (Tătui et al., 2016; Vidal-Ruiz and de Alegría-Arzaburu, 2019; and Tătui and Constantin, 2020).

Sonadia, especially the southeastern part of the island, undergoes dynamic morphological changes where there is a still developing sandbar (Hoque et al., 2013). The sandbar is protruding in the southeastern direction of the island. Besides, this active sandbar undergoes northward shifting. These sorts of morphological changes occur only in the southeastern portion of the island due to less exposure to open the Bay of Bengal and wave action than the western part encountering continuous wave action (Tătui et al., 2016; Franco-Ochoa et al., 2020; and Tătui and Constantin, 2020). Shore features like sandbars are very dynamic mainly governed by waves, currents, and regional sediment characteristics, and geological settings that may control the nearshore morphology by influencing the stability of sandbar morphology and position and surface sediment characteristics (McNinch, 2004). Similar findings were also supported by moderately sorted grain size distribution around Transect 3 and 4 compared to well-sorted to highly well-sorted sediment distribution across the rest of the shoreline of Sonadia Island (Hoque et al., 2013). These changes also may be attributed to longshore current around the island because the longshore current has an active role as waves generally do not approach the shoreline parallel to the shore which can move and distribute sediment. The second concept was also suggested by Ali (1989). He described discharge currents, tides, monsoon currents, and storm surges as the leading causes of coastal erosion or accretion, i.e., morphological changes. Beach drift may also have a contribution to the changes as it plays a role as waves approach at an angle to the beach, but retreat perpendicular to the shoreline. As a result, the swash of the incoming wave moves the sand up along the beach keeping a direction perpendicular to the incoming wave crests and the backwash moves the sand down the beach perpendicular to the shoreline (Fox et al., 1966). Consequently, the sand will move up and down along the beach following a zigzag path with successive waves (Ritter et al., 1995; Ahnert, 1996; Stephen, 1999). Over time, the effects could be enormous. Sediment is moved and re-deposited to increase the size of the shore of the island. The impact on the land surface was seen by examining the shore profile. Beaches occur where sand is deposited along the shoreline (Hayes, 1975). These physical forcings can provide extreme outputs as Moheshkhali Channel brings enormous sediment from the hilly region of Myanmar and Bangladesh (Islam and Hoque, 1999). Sediment is also transported here from the great GBM river system (Coleman, 1968; Ali, 1989 and Ali, 1991).

Conclusion

Morphological changes, especially erosion and deposition in the coastal area of Bangladesh, have taken place from time immemorial. We have some discontinuous records of coastal geomorphological changes of the last two hundred years, but they are mostly scanty. Records of study on morphological changes in Sonadia Island are not available. This study is the first one of its sort and very important, particularly in the context of the island developing as a new deep seaport. Therefore, a greater understanding of the hydro-geomorphology of the island system is critical for a mega-project like this. Further studies based on shoreline detection, nearshore satellite-derived bathymetry or the very recent automatic procedure to extract and analyze sandbar morphology and changes from satellite images should be conducted.

Acknowledgments

Special thanks go to the Global Land Cover Facility of the University of Maryland, USA, as Landsat satellite images have been acquired from its image archive. Thanks to the EROS (Earth Resources Observation System) Data Center's Distributed Active Archive Center (EDC-DAAC), a joint NASA-USGS data archive for the ASTER image and the Goggle Earth application interface for the QuickBird browse image.

References

Ahmad, Q.K., 2003. Regional cooperation in flood management in the Ganges-Brahmaputra-Meghna region: Bangladesh perspective. *Natural Hazards*, 28(1), pp. 191–198.

Ahnert, F., 1996. Introduction to geomorphology. *EOS Transactions American Geophysical Union* 81(40), pp. 463–463.

Akter, J., Sarker, M.H., Popescu, I. and Roelvink, D., 2016. Evolution of the Bengal Delta and its prevailing processes. *Journal of Coastal Research*, 32(5), pp. 1212–1226.

Aleman, N., Certain, R., Robin, N. and Barusseau, J.P., 2017. Morphodynamics of slightly oblique nearshore bars and their relationship with the cycle of net offshore migration. *Marine Geology*, 392, pp. 41–52.
Aleman, N., Robin, N., Certain, R., Anthony, E.J. and Barusseau, J.P., 2015. Longshore variability of beach states and bar types in a microtidal, storm-influenced, low-energy environment. *Geomorphology*, 241, pp. 175–191.
Ali, A., 1989. Storm surges in the Bay of Bengal and some related problems. Ph. D. thesis. University of Reading, England.
Ali, A., 1991. Satellite observation and numerical simulation of western boundary current in the Bay of Bengal. Presented in the 12th Asian Conference on Remote Sensing, Singapore, 30 October–5 November.
Allison, M. and Kepple, E., 2001. Modern sediment supply to the lower delta plain of the Ganges-Brahmaputra River in Bangladesh. *Geo-Marine Letters*, 21(2), pp. 66–74.
Almar, R., Castelle, B., Ruessink, B.G., Sénéchal, N., Bonneton, P. and Marieu, V., 2010. Two-and three-dimensional double-sandbar system behaviour under intense wave forcing and a meso–macro tidal range. *Continental Shelf Research*, 30(7), pp. 781–792.
Badrun, Y., 2017. Sandbar Formation in the Mesjid River Estuary, Rupat Strait, Riau Province, Indonesia. *The Indonesian Journal of Geography*, 49(1), p. 65.
Barua, D.K., 1991. The coastline of Bangladesh- An overview of Process and Forms. In: *Proceedings of 7th Symposium on Coastal and Ocean Management, USA*, pp. 2285–2301.
Bird, E.C.F., 1969. *Coasts: An Introduction to Systematic Geomorphology*. MIT Press, Massachusetts Institute of Technology, Massachusetts.
Bird, E.C.F., 2011. *Coastal Geomorphology: An Introduction*. John Wiley & Sons, New York, NY..
Biswash, A. K., 1978. Environmental implication of water development for developing countries. *Water Supply and Management*, 2, pp. 283–297.
Britannica, T., 2020. Editors of Encyclopedia. Sandbar. Encyclopedia Britannica. https://www.britannica.com/science/sandbar
Cohn, N., Ruggiero, P., Ortiz, J. and Walstra, D.J., 2014. Investigating the role of complex sandbar morphology on nearshore hydrodynamics. *Journal of Coastal Research*, 70(10070), pp. 53–58.
Coleman, J.M., 1968. The sediment field of major rivers of the world. *Water Resources Research*, 4(4), pp. 26–59.
Coleman, J.M., 1969. Brahmaputra River: Channel processes and sedimentation. *Sedimentary Geology*, 3, pp. 129–239.
Crawford, T.W., Rahman, M.K., Miah, M.G., Islam, M.R., Paul, B.K., Curtis, S. and Islam, M.S., 2021. Coupled adaptive cycles of shoreline change and households in deltaic Bangladesh: Analysis of a 30-year shoreline change record and recent population impacts. *Annals of the American Association of Geographers*, 111(4), pp. 1002–1024.
Curray, J.R. and Moore, D.G. 1971. Growth of the Bengal deep sea fan and denudation in the Himalayas. *Geological Society of America Bulletin*, 82, p. 563.
Datta, D.K. and Subramanian, V., 1997. Texture and mineralogy of sediments from the Ganges-Brahmaputra-Meghna river system in the Bengal Basin, Bangladesh and their environmental implications. *Environmental Geology*, 30(3), pp. 181–188.
Elgar, S., Gallagher, E.L. and Guza, R.T., 2001. Nearshore sandbar migration. *Journal of Geophysical Research: Oceans*, 106(C6), pp. 11623–11627.

Faria, A.G., Thornton, E.B., Lippmann, T.C. and Stanton, T.P., 2000. Undertow over a barred beach. *Journal of Geophysical Research: Oceans*, 105(C7), pp. 16999–17010.

Fein, J.S. and Stephens, P.L. (eds.), 1987. *Monsoons*. Wiley, New York, p. 631.

Fox, W. T., Ladd, J. W., and Martin, M. K., 1966. A profile of the four moment measures perpendicular to a shoreline, South Haven, Michigan. *Journal of Sedimentary Research*, 36(4), pp. 1126–1130.

Franco-Ochoa, C., Zambrano-Medina, Y., Plata-Rocha, W., Monjardín-Armenta, S., Rodríguez-Cueto, Y., Escudero, M. and Mendoza, E., 2020. Long-term analysis of wave climate and shoreline change along the Gulf of California. *Applied Sciences*, 10(23), p. 8719.

Gallagher, E.L., Elgar, S. and Guza, R.T., 1998. Observations of sand bar evolution on a natural beach. *Journal of Geophysical Research: Oceans*, 103(C2), pp. 3203–3215.

Ghosh, M.K., Kumar, L. and Roy, C., 2015. Monitoring the coastline change of Hatiya Island in Bangladesh using remote sensing techniques. *ISPRS Journal of Photogrammetry and Remote Sensing*, 101, pp. 137–144.

Ghosh, S., Guchhait, S.K., Illahi, R.A., Bera, S. and Roy, S., 2020. Geomorphic character and dynamics of gully morphology, erosion and management in laterite Terrain: Few observations from Dwarka–Brahmani Interfluve, Eastern India. *Geology, Ecology, and Landscapes*, pp. 1–29.

Grases, A., Gracia, V., García-León, M., Lin-Ye, J. and Sierra, J.P., 2020. Coastal flooding and erosion under a changing climate: Implications at a low-lying coast (Ebro Delta). *Water*, 12(2), p. 346.

Hayes, M.O., 1975. Morphology of sand accumulation in estuaries: An introduction to the symposium. In *Geology and Engineering* (pp. 3–22). Academic Press, Cambridge, MA.

Hoefel, F. and Elgar, S., 2003. Wave-induced sediment transport and sandbar migration. *Science*, 299(5614), pp. 1885–1887.

Hoque, M. E., Chowdhury, S. R., Uddin, M. M., Alam, M. S. and Monwar, M. M., 2013. Grain size analysis of a growing sand bar at Sonadia Island, Bangladesh. *Open Journal of Soil Science*, 3(2), pp. 71–80.

Hoque, M.E., Chowdhury, S.R., Chowdhury, M.Z.R. and Uddin, M.M., 2021. Morphological changes of a developing sandbar along the shoreline of Sonadia Island, Bangladesh between 1972 and 2006 using remote sensing. *Geology, Ecology, and Landscapes*, 5, pp. 1–9.

Hsu, T.J., Elgar, S. and Guza, R.T., 2006. Wave-induced sediment transport and onshore sandbar migration. *Coastal Engineering*, 53(10), pp. 817–824.

Islam M. R., Begum, S. F., Yamaguchi, Y. and Ogawa, K., 2002. Distribution of suspended sediment in the coastal sea off the Ganges–Brahmaputra River mouth: Observation from TM data. *Journal of Marine Systems*, 32(4), pp. 307–321.

Islam, A.S., Haque, A. and Bala, S.K., 2010. Hydrologic characteristics of floods in Ganges–Brahmaputra–Meghna (GBM) delta. *Natural Hazards*, 54(3), pp. 797–811.

Islam, M.A., Maitra, M., Majlis, A.K. and Rahman, S., 2012. Spatial changes of land use/land cover of Moheshkhali Island, Bangladesh: A fact-finding approach by remote sensing analysis. *Dhaka University Journal of Earth and Environmental Sciences*, 2, pp. 43–54.

Islam, M.A., Mitra, D., Dewan, A. and Akhter, S.H., 2016. Coastal multi-hazard vulnerability assessment along the Ganges deltaic coast of Bangladesh–A geospatial approach. *Ocean & Coastal Management*, 127, pp. 1–15.

Islam, M.M., Rahman, M.S., Kabir, M.A., Islam, M.N. and Chowdhury, R.M., 2020. Predictive assessment on landscape and coastal erosion of Bangladesh using geospatial techniques. *Remote Sensing Applications: Society and Environment*, 17, p. 100277.

Islam, M.S. and Hoque, A., 1999. Application of remote sensing technique to study the landuse changes of Moheshkhali Island in Bangladesh. *Journal of Remote Sensing and Environment*, 3, pp. 69–85.

Islam, N., 1991. Environmental Challenges to Bangladesh. Report. Bangladesh Institute of International and Strategic Studies, Dhaka, Bangladesh, Published.

Jackson, J.A., 2005. *Glossary of Geology* (5th edition). American Geological Institute, Alexandria, VA, p. 779.

Khan, M. S. H., Parkasha, B. T., Kumar, S., 2005. Soil–landform development of a part of the fold belt along the eastern coast of Bangladesh. *Geomorphology*, 71, pp. 310–327.

Khan, M.H.R., Liu, J., Liu, S., Seddique, A.A., Cao, L. and Rahman, A., 2019. Clay mineral compositions in surface sediments of the Ganges-Brahmaputra-Meghna river system of Bengal Basin, Bangladesh. *Marine Geology*, 412, pp. 27–36.

Khandker, H., 1997. Mean sea level in Bangladesh, *Marine Geodesy*, 20(1), pp. 69–76.

King, C.A.M. and Williams, W.W., 1949. The formation and movement of sand bars by wave action. *The Geographical Journal*, 113, pp. 70–85.

Kirwan, M.L. and Megonigal, J.P., 2013. Tidal wetland stability in the face of human impacts and sea-level rise. *Nature*, 504(7478), pp. 53–60.

Lafon, V., Apoluceno, D.D.M., Dupuis, H., Michel, D., Howa, H. and Froidefond, J.M., 2004. Morphodynamics of nearshore rhythmic sandbars in a mixed-energy environment (SW France): I. Mapping beach changes using visible satellite imagery. *Estuarine, Coastal and Shelf Science*, 61(2), pp. 289–299.

Lippmann, T.C. and Holman, R.A., 1990. The spatial and temporal variability of sand bar morphology. *Journal of Geophysical Research: Oceans*, 95(C7), pp. 11575–11590.

Masselink, G., Ruju, A., Conley, D., Turner, I., Ruessink, G., Matias, A., Thompson, C., Castelle, B., Puleo, J., Citerone, V. and Wolters, G., 2016. Large-scale Barrier Dynamics Experiment II (BARDEX II): Experimental design, instrumentation, test program, and data set. *Coastal Engineering*, 113, pp. 3–18.

McNinch, J.E., 2004. Geologic control in the nearshore: Shore-oblique sandbars and shoreline erosional hotspots, Mid-Atlantic Bight, USA. *Marine Geology*, 211(1–2), pp. 121–141.

Mikhailov, V.N. and Dotsenko, M.A., 2007. Processes of delta formation in the mouth area of the Ganges and Brahmaputra Rivers. *Water Resources*, 34(4), pp. 385–400.

Mukherjee, A., Fryar, A.E. and Thomas, W.A., 2009. Geologic, geomorphic and hydrologic framework and evolution of the Bengal basin, India and Bangladesh. *Journal of Asian Earth Sciences*, 34(3), pp. 227–244.

Mullick, M.R.A., Islam, K.A. and Tanim, A.H., 2019. Shoreline change assessment using geospatial tools: A study on the Ganges deltaic coast of Bangladesh. *Earth Science Informatics*, 13(2), 299–316.

Murty, T. S. and Henry R. F., 1983. Tides in the Bay of Bengal. *Journal of Geophysical Research*, 88, pp. 6069–6076.

Nelson, S.M., Zamora-Arroyo, F., Ramírez-Hernández, J. and Santiago-Serrano, E., 2013. Geomorphology of a recurring tidal sandbar in the estuary of the Colorado River, Mexico: Implications for restoration. *Ecological Engineering*, 59, pp. 121–133.

Nelson, S. 1999. *Lecture Notes Coastal Process.* Tulane University, New Orleans, LA.
Oyedotun, T.D.T., Ruiz-Luna, A. and Navarro-Hernández, A.G., 2018. Contemporary shoreline changes and consequences at a tropical coastal domain. *Geology, Ecology, and Landscapes,* 2(2), pp. 104–114.
Özpolat, E. and Demir, T., 2019. The spatiotemporal shoreline dynamics of a delta under natural and anthropogenic conditions from 1950 to 2018: A dramatic case from the Eastern Mediterranean. *Ocean & Coastal Management,* 180, p. 104910.
Pape, L., Plant, N.G. and Ruessink, B.G., 2010. On cross-shore migration and equilibrium states of nearshore sandbars. *Journal of Geophysical Research: Earth Surface,* 115(F3).
Pramanik, M.A.H., 1983. Remote Sensing applications to coastal morphological investigations in Bangladesh. Unpublished Ph. D. thesis. Department of Geography, Jahangirnagar University, Savar, Dhaka, Bangladesh.
Pramanik, M.A.H., 1988. Methodologies and techniques of studying coastal systems. SPARRSO case studies. Presented at the National Development and Management. Held in Dhaka, Bangladesh.
Pramanik, M.A.H., Ali, A. and Rahman, A. 1981. An assessment of the land accretion and erosion in the Meghna estuary. Presented at the third National Geographical Conference. Held in Dhaka, Bangladesh.
Rafati, Y., Hsu, T.J., Elgar, S., Raubenheimer, B., Quataert, E. and van Dongeren, A., 2021. Modeling the hydrodynamics and morphodynamics of sandbar migration events. *Coastal Engineering,* 166, p. 103885.
Ramage, C., 1971. *Monsoon Meteorology.* Academic Press, New York, p. 326.
Ritter, D.F., Kochel. RC and Miller, J.R. 1995. *Process Geomorphology* (3rd edition). WCB McGraw-Hill Publications, USA. pp. 1–544.
Rizzo, A. and Anfuso, G., 2020. Coastal dynamic and evolution: Case studies from different sites around the world. *Water,* 12(10), p. 2829.
Ruessink, B.G., Bell, P.S., Van Enckevort, I.M.J. and Aarninkhof, S.G.J., 2002. Nearshore bar crest location quantified from time-averaged X-band radar images. *Coastal Engineering,* 45(1), pp. 19–32.
Salam, M.I. and Hasan, K., 2019. Monitoring the coastline change of Moheshkhali Island using remote sensing techniques. *Journal of Spatial Hydrology,* 15(1), pp. 1–10.
Salauddin, M., Hossain, K.T., Tanim, I.A., Kabir, M.A. and Saddam, M.H., 2018. Modeling spatio-temporal shoreline shifting of a coastal island in Bangladesh using geospatial techniques and DSAS extension. *Annals of Valahia University of Targoviste, Geographical Series,* 18(1), pp. 1–13.
Sarwar, M. G. M., 2005. Impacts of Sea Level Rise on the Coastal Zone of Bangladesh, Master's thesis, Lund University International Masters Programme in Environmental Science, Lund University, Sweden, 45pp.
Sarwar, M.G.M. and Woodroffe, C.D., 2013. Rates of shoreline change along the coast of Bangladesh. *Journal of Coastal Conservation,* 17(3), pp. 515–526.
Siddiqui, M.H., 1988. Land accretion and erosion in the coastal area. Presented at the National Workshop on Bangladesh Coastal Area Resource Development and Management. Held in Dhaka, Bangladesh, 3–4 October, 1988.
SPARRSO Report, 1987. Report on pilot project on remote sensing application to coastal zone dynamics in Bangladesh. Dhaka, Bangladesh.
SPARRSO, 1993. Monitoring of changes in coastal zone area based on Landsat MSS and infrared aerial photographs, SPARRSO Report, December 1993.

Svendsen, I.A., 1984. Mass flux and undertow in a surf zone. *Coastal Engineering*, 8(4), pp. 347–365.

Tătui, F. and Constantin, S., 2020. Nearshore sandbars crest position dynamics analysed based on earth observation data. *Remote Sensing of Environment*, 237, p. 111555.

Tătui, F., Vespremeanu-Stroe, A. and Ruessink, G.B., 2016. Alongshore variability of cross-shore bar behavior on a nontidal beach. *Earth Surface Processes and Landforms*, 41(14), pp. 2085–2097.

Thornton, E.B., Humiston, R.T. and Birkemeier, W., 1996. Bar/trough generation on a natural beach. *Journal of Geophysical Research: Oceans*, 101(C5), pp. 12097–12110.

Vidal-Ruiz, J.A. and de Alegría-Arzaburu, A.R., 2019. Variability of sandbar morphometrics over three seasonal cycles on a single-barred beach. *Geomorphology*, 333, pp. 61–72.

Viles, H. and Spencer, T., 1995. *Coastal Problems*. Taylor & Francis, London, UK, p. 350.

Walstra, D.J.R., Reniers, A.J.H.M., Ranasinghe, R.W.M.R.J.B., Roelvink, J.A. and Ruessink, B.G., 2012. On bar growth and decay during interannual net offshore migration. *Coastal Engineering*, 60, pp. 190–200.

Walstra, D.J.R., Wesselman, D.A., Van der Deijl, E.C. and Ruessink, G., 2016. On the intersite variability in inter-annual nearshore sandbar cycles. *Journal of Marine Science and Engineering*, 4(1), p. 15.

Yuhi, M., Matsuyama, M. and Hayakawa, K., 2016. Sandbar migration and shoreline change on the Chirihama Coast, Japan. *Journal of Marine Science and Engineering*, 4(2), p. 40.

8

Integrated Soil Fertility Management for Climate Change Mitigation and Agricultural Sustainability

Tesfaye Bayu

CONTENTS

Introduction

Background and Justification

The world's population is estimated to reach 9.2 billion by 2050. Over this period, agricultural production must increase by 70% to keep pace with increasing food demand (FAO, 2000). More than 95% of global food comes

DOI: 10.1201/9781003175018-8

from land, so an adequate global food supply depends predominantly on the continued availability of productive soils. However, quality soils are not guaranteed without additional efforts. In addition, ongoing climate change (CC) has increased alterations of weather patterns, affected soil moisture availability, and brought associated consequences for diseases and pest incidences). By 2050, CC is expected to negatively impact at least 22% of the cultivated areas of the world's important crops, notably rice and wheat (Campbell and Campbell, 2011), and increase global warming. Global warming is caused by increased atmospheric concentrations of greenhouse gases (GHGs), mainly carbon dioxide (CO_2), methane (CH_4), and nitrous oxide (N_2O).

Agriculture is one of the largest contributors to GHG emissions, derived from livestock farming (e.g., enteric fermentation and manure management) and emissions from agricultural soils (i.e., application of excessive N fertilizers and decomposition of organic material). On average, agriculture accounts for about 14% of the total global GHG emissions (Parry et al., 2007). Contributing factors are poor land management by humans, such as over-cultivation, overgrazing and deforestation, draining of peatlands, and burning of rainforests.

Being part of the problem, agriculture is also part of the solution to CC impacts. If agricultural soils are properly managed and effective policies are in place, they have the potential to sequester large amounts of carbon from the atmosphere and store it in the soils, thereby mitigating CH_4 and CO_2 emissions (Gaskel et al., 2007).

Soil fertility and plant nutrition are important components of plant production. Productive capacity of soils requires the provision of adequate and balanced amounts of nutrients to ensure proper growth of the plants. The fact on the ground is that soil nutrient status of most farming systems is widely constrained by the limited use of inorganic and organic fertilizers and by nutrient loss mainly due to erosion and leaching (Tulema et al., 2007).

Nutrient management is one of the most important decision-making processes faced by those involved in the growth and production of plants for any purpose, whether it is as agronomic crops, as horticultural and landscape plantings in urban settings, or for the conservation and reclamation of disturbed lands.

Increasing the inputs of nutrients has played a major role in increasing the supply of food to a continually growing world population. However, over-application of inorganic fertilizers causes inefficient use, large losses, and imbalances of nutrients. It also leads to environmental contamination in several areas in the developed world. On the other hand, insufficient application of nutrients and poor soil management, along with harsh climatic conditions and other factors have contributed to the degradation of soils including soil fertility depletion in developing countries like Sub-Saharan Africa (SSA) (Goulding et al., 2008).

To boost crop production, farmers use both mineral and organic fertilizers to increase the condition of crop growth. The demerits of both mineral and organic fertilizers lead to the innovation of a new fertilizer called Organo-mineral fertilizers or integrated nutrient management. Many experiments have been conducted with the use of combined organic and mineral fertilizers for crop production in different formulations. Akande et al. (2010) combined kola pod husk with NPK fertilizer for production of Amaranthus. Ayeni et al. (2010) used combined poultry manure and NPK 20:10:10 fertilizer to increase the yield of maize and soil nutrients.

To replenish the soil nutrient depletion, application of chemical fertilizers is essential. However, high cost of chemical fertilizers coupled with the low affordability of small-holder farmers is the biggest obstacle for chemical fertilizer use. Moreover, the current energy crisis prevailing higher prices and lack of proper supply system of inorganic fertilizers calls for more efficient use of organic manure, green manure, crop residues, and other organic sources along with the inorganic fertilizers to sustain the yield levels (Sathish et al., 2011).

However, the application rate is often insufficient due to the low availability and high cost of mineral fertilizers. Further, problems with acidification may occur after intensive addition of ammonium-based N fertilizers (Vanlauwe and Giller, 2006). On the other hand, organic amendments show a slower nutrient release pattern than mineral fertilizer but facilitate an increased soil organic matter (SOM) content (Pinitpaitoon et al., 2011). Although Vanlauwe and Giller (2006) claim that organic resources are not sufficient to supply crops with the required nutrients, the increased SOM is enhancing productivity due to the improved biological activity and physical soil properties (Watson et al., 2002).

Continuous uses of inorganic fertilizers lead to deterioration of soil chemical and physical properties, biological activities, and thus in general the total soil health (Mahajan et al., 2008). Nutrients supplied exclusively through chemical sources, though enhance yield initially, lead to unsustainable productivity over the years (Satyanarayana et al., 2002; Mahajan et al., 2008). Thus, the negative impacts of chemical fertilizers, coupled with their high prices, have prompted the interest in the use of organic fertilizers as source of nutrients. Organic fertilizer application has been reported to improve crop growth by supplying plant nutrients including micronutrients as well as improving soil physical, chemical, and biological properties thereby provide a better environment for root development by improving the soil structure (Dejene et al., 2010).

Furthermore, the price of inorganic fertilizers is increasing and becoming unaffordable for resource-poor small-holder farmers. The best remedy for soil fertility management is, therefore, a combination of both inorganic and organic fertilizers, where the inorganic fertilizer provides readily available nutrients and the organic fertilizer mainly increases soil organic matter and improves soil structure and buffering capacity of the soil (Alemu, 2015). The combined application of inorganic and organic fertilizers, usually termed as integrated nutrient management, is widely recognized as a way of increasing yield and/or improving productivity of the soil sustainably (Mahajan et al.,

2008). Several researchers (Singh and Agarwal, 2001; Mahajan et al., 2008; Gafar et al., 2014) have demonstrated the beneficial effect of integrated nutrient management in mitigating the deficiency of several macro- and micronutrients. In view of this fact, identifying the optimum dose of integrated nutrients application is crucial and is required for maintaining adequate supply of nutrients for increased yield.

Integrated nutrient management (INM) is the combined use of mineral fertilizers with organic resources such as cattle manures, crop residues, urban/rural wastes, composts, green manures, and bio-fertilizers (Antil, 2012). Its basic concept is sustaining soil and crop productivity through optimization of all possible sources of plant nutrients in an integrated manner. In this system, all aspects of mineral and organic plant nutrient sources are integrated into the crop production system (FAO, 2006a, b) and are utilized in an efficient and judicious manner for sustainable crop production. It contributes to attaining agronomically feasible, economically viable, environmentally sound, and sustainable high crop yields in cropping systems by enhancing nutrient use efficiency and soil fertility, increasing carbon sequestration, reducing nitrogen losses due to nitrate leaching, and emission of greenhouse gases (FAO, 2006a, b; Milkha and Aulakh, 2010).

Integrated nutrient management implies the maintenance or adjustment of soil fertility and of plant nutrient supply to an optimum level for sustaining the desired crop productivity, on the one hand, and to minimize nutrient losses to the environment, on the other hand. It is achieved through efficient management of all nutrient sources. Nutrient sources to a plant growing on a soil include soil minerals and decomposing soil organic matter, mineral and synthetic fertilizers, animal manures and composts, by-products and wastes, plant residue, and biological N-fixation (Singh and Agarwal, 2001).

The diversity of agroecological zones across SSA (table) results in the wide range of farming systems. According to the availability of natural resources (land, water, grazing, areas, and forest) and climate, especially length of growing period and altitude, as well as the pattern of farm activities and household livelihood, African farming systems can be classified into different farming classes.

Objectives

General Objective

The overall objective of the chapter is to review on the contribution of integrated fertility management (IFM) to CC mitigation and agricultural sustainability

Specific Objectives

The specific objectives of the review were as follows:

- To review the contribution of IFM to CC mitigation
- To review the role of IFM for agricultural sustainability
- To review the residual advantage of IFM

Literature Review

Soil and CC soils are critical to food security but are too slowly formed and too quickly lost. Since climatic variables such as rainfall and temperature play an important role in the formation and/or destruction of soils (Brady and Weil, 2007), we need to better understand the impact of CC on soil processes and properties, and how soil management techniques contribute to CC adaptations/resilience, reduction in GHG emissions and increase in agricultural productivity. Soil resilience refers to the magnitude of disturbance (caused by CC in this case) that can be absorbed or accommodated before the system changes its structure (Seybold et al., 1999). The soil properties and functions that are closely related to soil resilience and mostly affected by CC are soil structure and texture, organic matter content, nutrient dynamics, soil organisms, soil pH, and cation exchange capacity (Figure 8.1).

Soils should be adequately monitored, protected, and maintained to ensure that the abovementioned crucial soil properties and functions remain in place. A range of soil management practices, including soil fertility improvements and soil erosion control, have been developed and applied by farmers and researchers in different parts of the world with a goal to achieve sustainable food security. However, a single soil management practice may solve part of the problem of CC impacts and food security, but not the whole problem. Understanding the status and condition of the soil properties is fundamental to making decisions to adopt or not to adopt soil management practices that contribute to climate-smart agriculture.

Climate-smart agriculture is based on the simultaneous achievements of three principal objectives:

i. adaptation to CC;
ii. mitigation of GHGs emissions; and
iii. increased agricultural productivity.

There is no universal definition adopted for Integrated Fertility Management (IFM). It all depends on the particular soil problem in the area. Therefore,

Property	*CC adaptation and mitigation potential*
Soil texture and structure	If soil structure is damaged by tillage operations during wet-dry conditions, the soil texture/structure functions to retain water and nutrients are hampered, and consequently the resilience to CC is reduced (FAO, 2013).
Soil organic matter (SOM)	SOM reduces evaporation, improves infiltration, increases soil aggregate stability; provides nutrients upon mineralization; and reduces surface runoff. These SOM functions increase the required adaptation potential to CC, build up resilience to water stress, and help mitigate GHG emissions (Giller *et al.*, 2009; FAO, 2013; Powlson *et al.*, 2011).
Soil nutrients and organisms	Need-based N management reduces losses of N and increases efficiency. Organisms living in soils (like earthworms) make nutrients available for uptake by plants through active involvement in the transformation and decomposition of nutrients., Soil organisms enhance adaptation to CC impacts, and increase crop production (Varinderpal-Singh *et al.*, 2012; Ali *et al.*, 2015).
Soil pH	Higher temperatures and evaporation rates increase accumulation of salts in the topsoil thereby raise the pH levels. Under such conditions, salinity is intensified, and water and nutrient availability is hampered, and soil structure is degraded. On the other hand, substantial increase in rainfall, accompanied by leaching, could lead to increase soil acidity. In consequence, the activity of microorganisms and nutrient release to crop plants is reduced, asnd does soil resilience to CC (FAO, 2013).

FIGURE 8.1
Carbon sequestration potential of soil. (a) Climate change (CC) adaptation and mitigation potential for soil texture and soil organic matter properties. (b) CC adaptation and mitigation potential for soil nutrients and soil pH.

there could be IFM for soil fertility improvements, integrated soil fertility management (ISFM) for soil erosion control, and so forth. According to Simpson et al. (2014), ISM for soil fertility improvements is a set of soil fertility management practices that entails the use of fertilizer, organic inputs, and improved germplasm combined with the knowledge on how to adapt these practices to local conditions, aiming at maximizing agronomic use efficiency of the applied nutrients and improving crop productivity (Figure 8.2).

More than 30 years of research on soil fertility, crop nutrition, and socioeconomics in small-holder farming systems of SSA has shown that combined interventions on fertilizer and organic inputs are prerequisites for achieving sustainable intensification. IFM builds on this notion and is originally defined as follows:

> A set of soil fertility management practices that necessarily includes the use of fertilizer, organic inputs, and improved germplasm combined with the knowledge on how to adapt these practices to local conditions in aim of maximizing the agronomic use efficiency of the applied nutrients

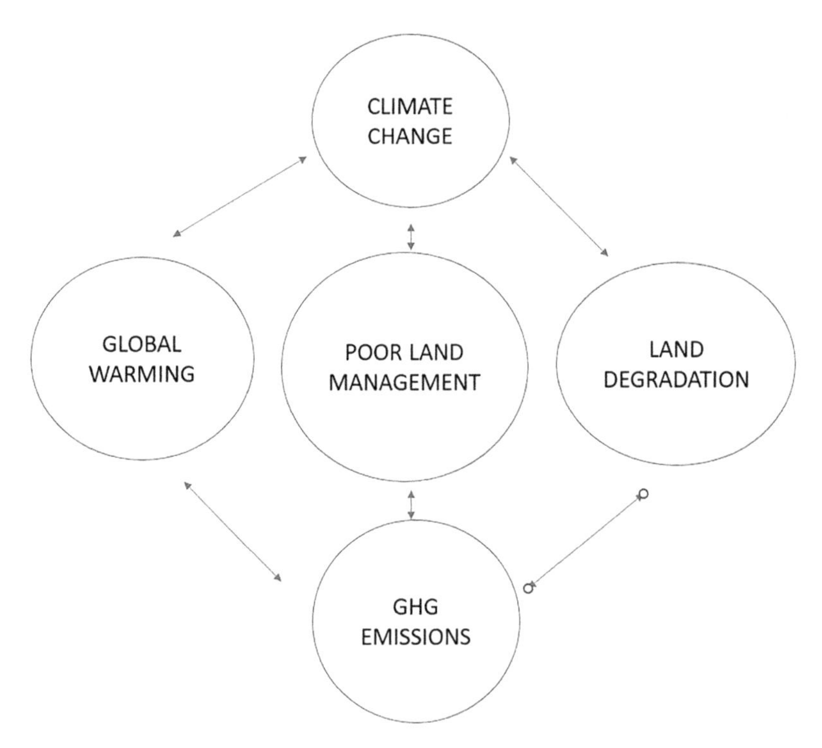

FIGURE 8.2
The relationship between climate change and soil management.

> and improving crop productivity. ISFM seeks that all inputs are managed following sound agronomic practices.
>
> *(Vanlauwe et al., 2010)*

Any of the interventions are required to increase the efficiency and profitability of food production as related to use of land, labor, fertilizer inputs, and financial investments

Adoption of ISFM, despite the significant benefits for food security, household income, and environmental protection, is challenged by the fact that adoption of practices of farmers is usually low and incomplete, especially in African small-holder systems.

The most important factors curtailing adoption are related to: (1) high transaction costs of input and produce trading (Alene et al., 2008), (2) low awareness and common disbeliefs about the benefits of soil fertility management (Schuijs et al., 2015), (3) shortage of credit facilities for making initial investments, (4) aversion to risks surrounding the profitability of inputs, (5) cost and availability of labor, (6) land size and property rights (Holden and Bezabih, 2008), (7) weak social networks and pervasive distrust (Wossen et al., 2015), (8) lack of information about soil fertility and rainfall forecasts (Maro et al., 2013), and (9) scarcity of organic residues and competition for residues with livestock (Giller et al., 2011).

Concepts of ISFM

Many paradigms of sustainable agriculture adhere to a combination of different and complementary agricultural technologies. Whether such a paradigm survives in practice depends on how farmers combine (or substitute) these technologies in their fields. Based on the work by Rauniyar and Goode (1992), we classify interrelationships in the application of different technologies by farmers into three main categories: independent, sequential, or simultaneous. Technologies are independent if the probability of application of one technology is not conditioned by the adoption of another technology. Sequential adoption takes place when the probability of application is conditioned on the adoption of another technology that precedes it. Finally, simultaneous adoption occurs when the probability of applying one technology is conditional on the adoption of another technology. The main biophysical rationale for farmers to combine different technologies is the existence of interaction effects on yield. Joint or sequential application of several technologies can have important non-linear effects, reducing or reinforcing the impact of a single technology on agricultural output, and/or leading to lasting effects on soil fertility and future productivity (Vanlauwe et al., 2010). For example, the agronomic efficiency of nitrogen in inorganic fertilizers is shown to significantly improve in combination with manure, and similarly, efficiency of nitrogen is significantly higher when applied on improved varieties (Vanlauwe et al., 2011).

ISFM is a means to increase crop productivity in a profitable and environmentally friendly way (Vanlauwe et al., 2011) and thus to eliminate one of the main factors that perpetuates rural poverty and natural resource degradation in SSA. Current interest in ISFM partly results from widespread demonstration of the benefits of typical ISFM interventions at plot scale, including the combined use of organic manure and mineral fertilizers (Zingore et al., 2008), dual-purpose legume–cereal rotations (Sanginga et al., 2003), or micro-dosing of fertilizer and manure for cereals in semiarid areas (Tabo et al., 2007). ISFM is also aligned to the principles of sustainable intensification (Pretty et al., 2011), one of the paradigms guiding initiatives to increase the productivity of small-holder farming systems. Sustainable intensification, though lacking a universally accepted definition, usually comprises aspects of enhanced crop productivity, maintenance and/or restoration of other ecosystems services, and enhanced resilience to shocks. ISFM can increase crop productivity and likely enhances other ecosystems services and resilience by diversifying farming systems, mainly with legumes, and increasing the availability of organic resources within farms, mainly as crop residues and/or farmyard manure.

One of the principles of ISFM—the combined application of fertilizer and organic resources—has been promoted since the late 1980s (Vanlauwe et al., 2001a), because of (1) the failure of Green Revolution-like interventions in SSA and (2) the lack of adoption of low-external-input technologies by small-holder farmers, including herbaceous legume-based technologies. The combined application of fertilizer and organic inputs made sense since (1) both fertilizer and organic inputs are often in short supply in small-holder farming systems due to limited affordability and/or accessibility; (2) both inputs contain varying combinations of nutrients and/or carbon, thus addressing different soil fertility-related constraints; and (3) extra crop produce can often be observed due to positive direct or indirect interactions between fertilizer and organic inputs (Vanlauwe et al., 2001a). In 1994, Sanchez (1994) presented the "second paradigm" for tropical soil fertility management, to "overcome soil constraints by relying on biological processes by adapting germplasm to adverse soil conditions, enhancing soil biological activity, and optimizing nutrient cycling to minimize external inputs and maximize their use efficiency." In this context, he already highlighted the need to integrate improved germplasm, a second principle of ISFM, within any improved strategy for nutrient management.

ISFM aims at the optimal and sustainable use of soil nutrient reserves, mineral fertilizers, and organic amendments as well as improved germplasm. Combining increases crop yield, rebuilds depleted soils, protects the natural resource base, and focuses on application of locally adapted SFM practices (Vanlauwe et al., 2001a).

Principles of ISFM

Maximize use of organic materials: Organic inputs (crop residues and animal manures) are also an important source of nutrients, but their N, P, Mg, and Ca content is only released following decomposition. By contrast, K is released rapidly from animal manures and crop residues because it is contained in the cell sap. Further, the amount of nutrients contained in organic resources is usually insufficient to sustain required levels of crop productivity and realize the full economic potential of a farmer's land and labor resources (Alun, 2020).

Judicious use of inorganic fertilizer: Mineral fertilizers are required to supplement the nutrients recycled or added in the form of crop residues and animal manures. Fertilizers are concentrated sources of essential nutrients in a form that is readily available for plant uptake. They are often less costly than animal manures in terms of the cost of the nutrients that they contain (i.e., $/kg nutrient) but often viewed as more costly by farmers because they require a cash outlay (Alun, 2020).

Use of improved germplasm: It is important that the farmer uses the crop planting materials (usually seed but sometimes seedlings) best adapted to the particular farm in terms of (Alun, 2020):

- Responsiveness to nutrients (varieties differ in their responsiveness to added nutrients);
- Adaptation to the local environment (soils, climate); and
- Resistance to pests and diseases (unhealthy plants do not take up nutrients efficiently).

Effect of ISFM for Soil Fertility Improvement

Soil fertility can be defined as the capacity of soil to provide physical, chemical, and biological needs for the growth of plants for productivity, reproduction, and quality, relevant to plant and soil type, land use, and climatic conditions (Abbott and Murphy, 2007). It is becoming understandable that the proper agricultural use of soil resources requires equal consideration for biological, chemical, and physical components of soil. Soil fertility is thus attaining a sustainable agricultural system.

The first step in maintaining soil fertility should be directed at maintaining the organic matter content of the soil. This can be done by using appropriate crop husbandry practices and by applying organic manure or compost together with mineral fertilizer. Chemical fertilizers can restore the soil fertility very quickly whereas organic fertilizers will provide nutrients to the soil in slow way (Van Scholl and Nieuwenhuis, 2007).

It is generally known that the incorporation of fertilizers is increasing yield and agricultural productivity. The combination of both organic and mineral fertilizers is crucial as they influence different soil properties. Mineral fertilizers are characterized by a high concentration of plant-available nutrients. Several studies showed a significant increase in grain yield after mineral fertilizer treatment. Drechsel et al. (2001) claim that fertilizer application is increasing with increasing population pressure at small-holder level. At small-holder level, organic material is applied in the form of farmyard manure (FYM) as it is often the source of organic matter (Dunjana et al., 2012).

Residual Advantage of Integrated Soil Fertility Management

Reviewing the residues of fertilizers on succeeding crops, Cooke (1970) reported that past manuring with farmyard manure and fertilizers leaves residues of nitrogen, phosphorus, and potassium in soil that benefit the following crops. He further indicated that the residues of inorganic nitrogen fertilizers usually last only for a season, but the residual effects of continued manuring with phosphorus and potassium may last for many years.

Akande et al. (2003) also reported an increase in soil available P of between 112% and 115% and 144% and 153%, respectively, for a 2-year field trial, after applying rock phosphate with poultry manure on okra. Akande et al. (2005) further reviewed the effect of rock phosphate amended with poultry manure on the growth and yield of maize and cowpea and reported that when rock phosphate application had continued over a period of several years, a large pool of undissolved rock phosphate could accumulate.

Residual effects of manure or compost application can maintain crop yield level for several years after manure or compost application ceases since only a fraction of the N and other nutrients in manure or compost become plant available in the first year after application (Eghball, 2002). Eghball and Power (1999) found that 40% of beef cattle feedlot manure N and 20% of compost N became plant available in the first year after application, indicating that about 60% of manure N and 80% of compost N became plant available in the succeeding years, assuming little or no loss of N due to NO_3^- leaching or denitrification. Residual effects of organic materials on soil properties can contribute to improvement in soil quality for several years after application ceases (Ginting et al., 2003).

Cooke (1970) found that 184.8 kg N/ha given to potatoes raised yields of wheat the following year which received no fresh fertilizer nitrogen from 3463.8 to 4570.5 kg/ha, but even where the wheat received a fresh dressing of 123.2 kg N/ha residues from the dressing given to the previous potatoes still raised yields by 764.5 kg/ha. Further results showed that when soil contains residues of inorganic nitrogen, larger maximum yields are possible than may be obtained from soil without residues. The results also showed that dressings of inorganic N fertilizers had large residual effects in the first year after the dressings stopped but much smaller effects in the second and third years.

Manure fertilizer treatments had beneficial residual effects on crop production and use from manure fertilizer for field fertilization and production of crops was better improved. Significantly high grain yield was obtained from residual application of 8 t/ha and is proportional with existing fertilizer recommendation. Therefore, for resource-poor farmers combined application of farmyard manure and mineral fertilizer is very economical than sole NP application (Chekole, 2015).

A study conducted at Ethiopia using nug as proceeding crop indicated that maize grain yields were significantly increased in rotation with this crop

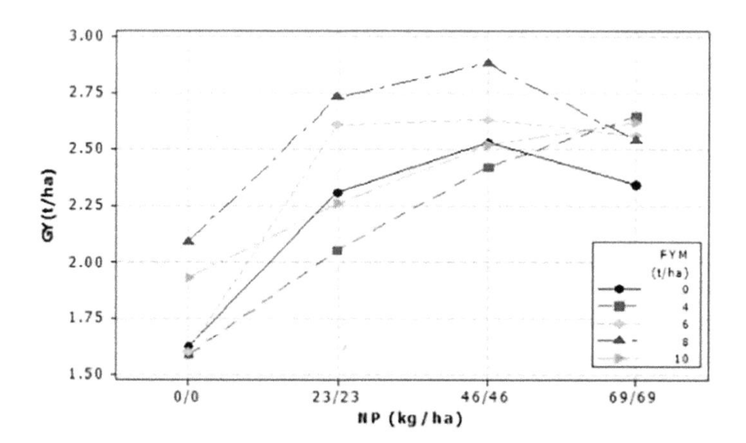

FIGURE 8.3
Residual effect of farm yard manure on grain of barley. (a) Ordinate in in GY (t/ha) from 1.50 to 3.00 and abscissa is in NP (kg/ha) from 0/0 to 69/69.

compared to the continuous cropped maize (Figure 8.3). This result clearly demonstrated the residual benefits of crop rotation with reduced NP fertilizer amendments and enhanced maize grain yield. Also, the integrated use of precursor crops with low rate of NP and farmyard manure gave comparable maize yield to a plot received recommended fertilizer rate (110/20 kg NP ha^{-1}). Production of maize following nug as a precursor crop by integrating with 46/5 kg ha^{-1} NP and 8 t FYM ha^{-1} could be affordable for small-holder farmers in Ethiopia areas (Berhanu, 1985). Means with the same letter in the same column are not significantly different at 5% using Duncan's Multiple Range Test (DMRT).

The possible reason for maximum height in FYM or VC (vermicompost) plus mineral NP treatment might be that the mineral NP sources fulfilled the NP requirements at early growth stages while farmyard manure and vermicompost provided the crop with maximum nutrients in later stages.

Thus, combination of (FYM+inorganic NP and VC+inorganic NP) might have nourished the crop in initial stages as well as in the later growth stages. The result of this experiment agreed with the finding of Amanuliah and Maimoona (2007) who reported that the use of increased rates of FYM and N increased plant height of wheat and the shortest plants recorded from the control treatment. Also, in agreement with this result, Ofosu and Leitch (2009) reported that plant height of spring barley increased with organic manure application as compared to inorganic fertilizer alone. Similarly, Getachew reported that the use of organic manures in combination with mineral fertilizers maximized the plant height than the application of inorganic fertilizers alone (Table 8.1).

Generally, it was observed that except the combined application of 2.5 t/ha VC with 25% and 50% inorganic NP fertilizers both at Adiyo and Ghimbo,

TABLE 8.1

Major Farming Systems in Sub-Saharan Africa

Farming System	Percent of Land	Principal Crop
Integrated	9	Rice, cotton, vegetables, rainfed crops, cattle, poultry
Tree crop	18	Cocoa, coffee, oil palm, rubber, yams, maize, off-farm work
Forest based	14	Cassava, maize, beans, cocoa, yams
Maize mixed	33	Maize, tobacco, cotton, cattle, goats, poultry, off-farm work
Agro pastoral	8	Sorghum, pearl millet, pulses, sesame, cattle, sheep, goats, poultry, off-farm work
Pastoral	17	Cattle, camels, sheep, goats, remittances
Urban based	1	Fruit, vegetables, dairy, cattle

the combined application of organic and inorganic fertilizers has resulted in higher aboveground biomass yield than the application of 100% recommended rate of inorganic NP alone. This implies that integrated use of organic and inorganic fertilizers responded better to increase productivity than the use of inorganic fertilizer alone in the study areas. Likewise, Shata et al. (2007) suggested that using mixed chemical and bio-fertilizers not only production can be kept at optimum level, but also the amount of chemical fertilizer to be used can be reduced. Plant biochemical activities improve by absorption of nutrients from soil, and this increases the grain yield and biological yield plant^{-1}.

Research on wheat and tef revealed that the application of different soil fertility management treatments significantly ($p<0.05$ and $p<0.01$) affected organic carbon, total N, available P, nitrate N (NO_3-N), and ammonium N (NH_4-N) analyzed for samples taken after harvesting from trial fields of both crops. Soil pH of wheat fields was significantly ($p<0.05$) affected by different soil fertility management treatments, but not soil pH of tef trial sites (Tables 8.2 and 8.3). Different soil fertility management treatments had significant effects on post-harvest soil organic carbon content. A significant improvement was observed in organic carbon content compared to the contents of the soil before treatment application. Relatively higher soil organic carbon was recorded on experimental plots, which received either organic or inorganic and organic nutrient sources (Tables 8.2 and 8.3) than plots that received only inorganic fertilizers (Agegnehu et al., 2014).

Effect Integrated Soil Fertility Management on Climate Mitigation

Healthy soils provide the largest store of terrestrial carbon. When managed sustainably, soils can play an important role in CC mitigation by storing

TABLE 8.2

Maize Productivity Along Different Treatment

Plant Height (cm)	No. of Leaves (cm^2)	Leaf Area (t/ha)	Stover Yield (t/ha)	Grain Yield (t/ha)	Root Dry (%)	Matter	Increase in Grain
Control	72.60	8.00	14	3.23	2.84	0.67	-
–2.5 t/ha OG	89.70	9.33	20	3.59	3.00	0.93	5.63
5 t/ha OG	107.90	9.23	19	3.97	3.11	0.97	9.51
10 t/ha OG	149.40	12.00	32	4.99	4.25	0.99	49.65
2.5 t/ha OMF	129.40	12.20	44	5.34	4.55	1.10	60.21
5 t/ha OMF	169.20	14.59	31	5.36	4.78	1.00	68.31
10 t/ha OMF	164.10	12.40	30	4.63	3.94	0.97	38.72
300 kg/ha NPK	194.00	12.3	24	4.23	3.44	0.93	12.13

Source: Adapted from Vanlauwe, unpublished data.

TABLE 8.3

Effect of Integrated Soil Fertility Management on Soil Properties

Treatments (kg/ha)	pH (H_2O)	OC (%)	N (%)	P (ppm)	NO_3^- (ppm)	NH_4^+ (ppm)
Control	5.57	1.36	0.14	9.4	6.00	8.55b
Farmers NP rate (23/10/0)	5.36	1.61	0.16	11.00	6.33b	9.25
Recommended NP rate (60/20/0)	5.26	1.83	0.17	15.55	7.20	9.78
50% of recommended NP rate+50% manure+50% of compost as N equivalence	5.76	2.06	0.18	15.57	10.60	13.60
50% of manure+50% of compost as N equivalence	6.15	1.98	0.17	15.52	9.78	10.70
F-probability	*	**	*	**	**	*
$LSD_{0.05}$	0.39	0.21	0.02	3.40	1.82	2.97
CV (%)	4.55	13.2	2.69	16.40	14.81	18.61

*, **=significant at $p<0.05$ and $p<0.01$, respectively.

carbon (carbon sequestration) and decreasing greenhouse gas emissions in the atmosphere. Conversely, if soils are managed poorly or cultivated through unsustainable agricultural practices, soil carbon can be released into the atmosphere in the form of carbon dioxide (CO_2), which can contribute to CC. The steady conversion of grassland and forestland to cropland and grazing lands over the past several centuries has resulted in historic losses of soil carbon worldwide. However, by restoring degraded soils and adopting soil conservation practices, there is major potential to decrease the emission of greenhouse gases from agriculture, enhance carbon sequestration, and build resilience to CC (FAO and ITPS, 2015).

Soil hosts the largest terrestrial carbon pool, and the biogeochemical processes that take place in the soil regulate the exchange of greenhouse gases with the atmosphere (Scharlemann et al., 2014). These processes and emissions are strongly affected by land use, land-use change, vegetation cover, and soil management. The stocks of soil organic carbon in the upper soil layers are particularly responsive to these influences, and their careful management provides an opportunity to reduce the concentration of greenhouse gases in the atmosphere.

Sustainable soil and land management interventions that are designed to increase soil organic matter should be accompanied by actions that address the drivers of degradation and help preserve existing soil carbon stocks, particularly in soils with high soil organic carbon content (Smith et al., 2014). Carbon sequestration in soils will contribute to CC adaptation and mitigation. It will also make agricultural production systems more sustainable; increase the overall resilience of agricultural ecosystems; and maintain the ecosystem services that are supported by soils (FAO, 2006a, b).

Healthy soils provide the largest store of terrestrial carbon. When managed sustainably, soils can play an important role in CC mitigation by storing carbon (carbon sequestration) and decreasing greenhouse gas emissions in the atmosphere. Conversely, if soils are managed poorly or cultivated through unsustainable agricultural practices, soil carbon can be released into the atmosphere in the form of carbon dioxide (CO_2), which can contribute to CC. The steady conversion of grassland and forestland to cropland and grazing lands over the past several centuries has resulted in historic losses of soil carbon worldwide. However, by restoring degraded soils and adopting soil conservation practices, through IFM there is major potential to decrease CC (Sainju et al., 2008).

A substantial amount of global CO_2 comes from soil through decomposition, mineralization, and soil respiration. So, when fertilizers were added to the soil through integrated way the decomposition rate was reduced and carbon dioxide emission was altered.

Nutrient management strives to balance the withdrawal of soil nutrients from fields, pastures and orchards by crops, livestock, and natural processes with the addition of nutrients provided by crop residues, compost, manure, or commercial fertilizers. The main objective of nutrient management is to optimize the yield and quality of crop production, while minimizing costs and negative environmental impacts. Failure to properly manage nutrients results in poor nutrient use efficiency and potentially harmful downstream environmental effects. Good nutrient management prevents the over-application of essential crop nutrients and sustainable nutrient management considers the full cost associated with application, including the energy embedded in added nutrients (Allen, 2011).

CC mitigation involves reducing the amount of greenhouse gases in the atmosphere or enhancing their sinks, e.g., by reducing the use of fossil fuels, planting trees, or enhancing mineralization of organic matter into soil organic carbon (John et al., 2014).

Adopting the INM strategy is essential to soil organic carbon (SOC) sequestration. The sink capacity of SOM for atmospheric CO_2 is greatly enhanced when soils are treated with integrated nutrient management instead of treating it with organic or inorganic source of nutrient (Lal, 2004a).

World cropland soils cover about 1.5 b ha and have a large capacity to sink carbon (Lal, 2010). Management of soil organic carbon pool is an important aim to achieve adaptation to and mitigation of global CC, while advancing global food security (Lal, 2004b). As an important sink of carbon, cropland soils can be used to mitigate and adapt to global CC. The rate and total magnitude of soil organic carbon sequestration (an average of about 0.55×10^{-9} Pg C ha^{-1} y^{-1}) depend on residue management and recycling of organics, climate regime, N application, and soil properties. Like cropland soils, forest and grassland soils can also be important for carbon sequestration.

Many factors are involved in carbon sequestration in forest soils, including carbon input by litter and roots into different soil horizons, soil age, N application, moisture regime, site management, frequency, intensity of burning, the addition of charcoal, and residue management (Lal, 2005a, 2005b). McKinsey and Company estimated that by 2030, afforestation can mitigate 0.27 Pg C/yr; reforestation, 0.38 Pg C/yr; and improved management, 0.08 Pg C/yr. Grassland soils cover 2.9 b ha globally, including 2.0 b ha under tropical grasslands or savannas and 0.9 b ha under temperate grasslands (Lal, 2010). Possible management practices for C sequestration in grassland can be fertilization, controlled grazing, conversion of degraded cropland and native vegetation to pasture, sowing of leguminous and grass pasture species, fire management, and water conservation. Mean rate of soil C sequestration in grassland is 5.4×10^{-10} Pg C/ha/yr.

Improving Nitrogen-Use Efficiency

The most effective method for reducing N_2O emissions is to increase nitrogen-use efficiency by applying precise amounts of nitrogenous fertilizer with manure to crops based on N estimates from soil and plant tissue tests.

Precisely timing N fertilizer applications will also increase nitrogen-use efficiency, ultimately leaving less N in the soil available for microbes to break down and release as N_2O. Accurate timing will also reduce fertilizer N losses due to nitrate (NO_3) leaching (FAO, 2010).

INM reduces emission of GHG by the following:

- Use recommended rates of suitable organic and inorganic fertilizers.
- Place the nitrogen more precisely into the root zone to make it more accessible by crops.
- If possible, use precision agriculture techniques to improve fertilizer application by helping determine exactly where to place nutrients, how much to apply, and when to apply. Three techniques can help achieve this objective.
- The collection of spatial data from pre-existing conditions in the field (e.g., remote sensing, canopy size, or yield measurement).
- The application of precise fertilizer amounts to the crop when and where needed.
- The recording of detailed logs of all fertilizer applications for spatial and temporal mapping.

Improvement in soil fertility through nutrient management is also important to SOC sequestration (Lal, 2005c) because concentrations of SOC and N are key indicators of soil quality and productivity through their favorable effects on physical, chemical, and biological processes, including nutrient cycling, water retention, root and shoot growth, and environmental quality (Sainju and Good, 1993).

Effect of Integrated Soil Fertility Management on Agricultural Sustainability

The efficiency of applied chemical fertilizers is also increased when applied along with organic manures. Therefore, better management of soil nutrients is required that delivers sustainable agriculture and maintains the necessary increases in food production while minimizing waste, economic loss, and environmental impacts (Goulding et al., 2008). Various long-term research results have shown that neither organic nor mineral fertilizers alone can achieve sustainability in crop production. Rather, integrated use of organic and mineral fertilizers has become more effective in maintaining higher productivity and stability through correction of deficiencies of primary, secondary, and micronutrients (Milkha and Aulakh, 2010). Therefore, judicious use

FIGURE 8.4
Effect of INM on sustainable crop growth.

of integrated nutrient management is the best alternative to supply nutrient to crop needs and improve soil conditions (Naresh et al., 2013).

So far, many research findings have shown that neither inorganic fertilizers nor organic sources alone can result in sustainable productivity (Satyanarayana et al., 2002).

For sustainable crop production, integrated use of chemical and organic fertilizer has proved to be highly beneficial. Several researchers have demonstrated the beneficial effect of combined use of chemical and organic fertilizers to mitigate the deficiency of many secondary and micronutrients in fields that continuously received only N, P, and K fertilizers for a few years, without any micronutrient or organic fertilizer (Figure 8.4). Research has shown that combinations of organic and mineral fertilizers result in greater crop yields compared with sole organic or sole mineral fertilizers (Chivenge et al., 2009). Vanlauwe et al. (2002) reported that grain yield increases of up to 400% over the control in cases where the control yields are low. This increase in grain yield has been attributed to improved N synchrony with combined inputs through direct interactions of the organic and N fertilizers.

Effect of Integrated Soil Fertility Management on Crop Productivity

The weakness in the productivity of crops across SSA is not only related to the poor soils in many countries (Chekole, 2015) but also to the limited use of essential inputs that are needed to raise the productivity level. These inputs include

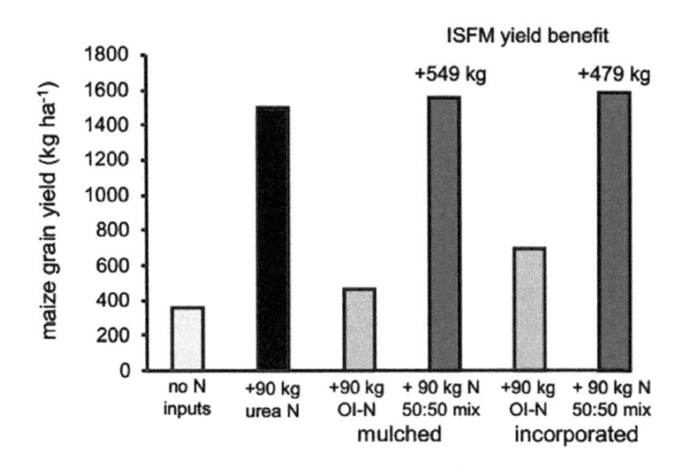

FIGURE 8.5
Interaction between mineral fertilizer and farmer available organic input (OI).

the use of improved seeds, fertilizers, irrigation, and pesticides. The hypothesis is that the use of these inputs would boost the productivity of crops (Alun, 2020). Research results indicated that productivity of wheat was significantly affected by different soil fertility treatment levels. Applications of inorganic and organic nutrient sources either alone or in combination had a significant ($p<0.001$ and $p<0.01$) effect on grain yield, total biomass, and harvest index of wheat, but not on its thousand grain weight (Figure 8.5). Analysis of variance over 2 years indicated that the year by soil fertility treatment level interaction (YxT) effect was significant ($p<0.05$ and $p<0.001$) for wheat grain yield, total biomass, and harvest index, but not for thousand grain weight (Agegnehu et al., 2016).

Effect of Integrated Soil Fertility Management on Environmental Sustainability

The soil sustains most living organisms, being the ultimate source of their mineral nutrients. Good management of soils ensures that mineral elements do not become deficient or toxic to plants, and that appropriate mineral elements enter the food chain. Soil management is important, both directly and indirectly, to crop productivity, environmental sustainability, and human health. Because of the projected increase in world population and the consequent necessity for the intensification of food production, the management of soils will become increasingly important in the coming years. To achieve future food security, the management of soils in a sustainable manner will be the challenge, through proper nutrient management and appropriate soil conservation practices. Research will be required to avoid further degradation of soils, through erosion or contamination, and to produce sufficient safe and nutritious food for healthy diets (Philip et al., 2012).

Long-term sustainability of agro-ecosystems requires soil protection from degradation and reduction of greenhouse gas emissions and of environmental pollution. Soil protection needs judicious and prudent use of conservation agriculture to prove its potential as a conservation effective technology, climate-resilient agriculture, and a viable option for sustainable intensification of agro-ecosystems for advancing food security and for adaptation to/mitigation of CC. Conservation agriculture refers to a farming system comprised of crop residue mulch, cover cropping, integrated nutrient management (INM), and no tillage techniques in a rotation cycle for effective soil and water conservation, carbon sequestration, sustainable intensification, and CC adaptation and mitigation (Lal, 2015).

Soils host a huge biodiversity of microbes and fauna which is not yet well understood: the small size of the soil-borne organisms; their immense diversity; the difficulty in isolating them; and the great heterogeneity of their habitats across different scales. The soil biodiversity studies include microbes (archaea, bacteria, fungi) and fauna (protozoa, micro-arthropods, nematodes, oligochaeta), and their relationship with aboveground biodiversity. We need to extend our capability to explore biological dynamics of soils at the scientific level, increase our knowledge of soil biodiversity and its role in ecosystem services across different soils, climate types and land use at the technological level, standardize methods and operating procedures for characterizing soil biodiversity and functioning, and develop bio-indicators at the economic level, assess the added value brought by cost-effective bio-indicators, and cost-effectiveness of alternative ecosystem services maintenance policies. For improving soil biological properties, we need to deploy our efforts with three approaches: description of soil biodiversity and of the relations between soil biodiversity, soil functions, and ecosystem services; long-term observatories representative of soil types, climates, and land uses, and modeling to elucidate relationships between soil biodiversity and functions (Jahangir et al., 2018).

Conclusions

From the review of different literature, the following points are concluded:

- It is possible to conclude INM improves yield and yield components for different crops.
- INM is more advantageous than other soil fertility management methods due to residual nutrients that help to produce crop more than one season.

- INM is also important to mitigate CC by increasing carbon sequestration and increasing N-use efficiency.
- Additionally, INM plays a role in substantiality of agricultural productivity and soil fertility.
- Nowadays INM is becoming a soil fertility and yield improving practice in some parts of Ethiopia.

Personal Arguments

Based on the review, the following are my arguments:

- INM is best in its role on soil fertility and yield improvement in sustain way.
- Especially for countries like Ethiopia integrating all nutrient source is very good to minimize the economic pressure on household income.
- It is also very good b/c of its residual effect.

Acknowledgment

Part of this document and some figures are from "Bayu, T. 2020. Review on contribution of integrated soil fertility management for climate change mitigation and agricultural sustainability. *Cogent Environmental Science*, 6, 1823631," with permission of Taylor & Francis Group.

References

Abbott, L. K., and Murphy, D. V. 2007. *Soil Biological Fertility – A Key to Sustainable Land Use in Agriculture*. Springer, Dordrecht, NL.

Agegnehu, G., Bass, A. M., Nelson, P. N., and Bird, M. I. 2016. Benefits of biochar, compost and biochar– compost for soil quality, maize yield and greenhouse gas emissions in a tropical agricultural soil. *Science of the Total Environment*, 543, 295–306

Agegnehu, G., vanBeek, C., and Bird, M. I., Agegnehu. 2014. Influence of integrated soil fertility management in wheat and tef productivity and soil chemical properties in the highland tropical environment. *Journal of Soil Science and Plant Nutrition*, 14(3), 532–545.

Akande, M. O., Adediran, J. A., and Oluwatoyinbo, F. I. 2005. Effects of rock phosphate amended with poultry manure on soil available P and yield of maize and cowpea. *African Journal of Biotechnology*, 4, 444–448.

Akande, M. O., Oluwatoyinbo, F. I., Adediran, J. A., Buari, K. W., and Yusuf, I. O. 2003. Soil amendments affect the Release of P from rock phosphate and the development and Yield of Okra. *Journal Vegetable Crop Production*, 19, 3–9.

Akande, M. O., Oluwatoyinbo, F. I., Makinde, E. A., Adepoju, A. S., and Adepoju, I. S. 2010. Response of okra to organic and inorganic fertilization. *Nature and Science*, 8(11), 261–266.

Alemu, M. M. 2015. Effect of tree shade on coffee crop production. *Journal of Sustainable Development*, 8(9), 66. https://doi.org/10.5539/jsd.v8n9p66

Alene, A. D., Manyong, V. M., Omanya, G., Mignouna, H. D., Bokanga, M., and Odhiambo, G. 2008. Smallholder market participation under transactions costs: Maize supply and fertilizer demand in Kenya. *Food Policy*, 33(4), 318–328

Allen, D. 2011. *What is Soil Biological Fertility? Farm Practices and Climate Change Adaptation Series*. Agriculture and Agri-Food Canada and BC Ministry of Agriculture, Victoria, BC.

Alun, T. 2020. Improving Crop Yields in Sub-Saharan Africa: What Does the East African Data Say? IMF Working Papers, Washington, DC.

Amanuliah, J., and Maimoona, N. 2007. The response of wheat (*Triticum aestivum* L.) to farm yard manure and nitrogen under rain fed condition. *African Crop Science Proceeding*, 8, 37–40.

Antil, R. S. 2012. Integrated plant nutrient supply for sustainable soil health and crop productivity. vol. 3. Focus Global Reporter.

Ayeni, J. M., Kayode, J., and Tedela, P. O. 2010. Allelopathic potentials of some crop residues on the germination and growth of *Bidens pilosa* L. *Nong Ye Ke Xue Yu Ji Shu*, 4(1), 21.

Berhanu, D. 1985. The Vertisols of Ethiopia, their characteristics, classification and management. In *Fifth Meeting of the Eastern African Sub-Committee for Soil Correlation and Land Evaluation, Wad Medani, Sudan, 5–10 December 1983* (pp. 31–54). World Soil Resources Report No. 56. FAO, Food and Agriculture Organization, Rome.

Brady, N. C., and Weil, R. R. 2007. Micronutrients and other trace elements. In *The Nature and Properties of Soils* 13th Edition, Prentice Hall, Upper Saddle River, NJ (pp. 654–684).

Campbell, S. G., and Campbell, K. S. 2011. Mechanisms of residual force enhancement in skeletal muscle: Insights from experiments and mathematical models. *Biophysical Reviews*, 3(4), 199–207.

Chekole, A.W. 2015. Response of Barley (*Hordium vulgare* L.) to integrated cattle manure and mineral fertilizer application in the vertisol areas of South Tigray, Ethiopia. *Journal of Plant Sciences*, 3(2), 71–76.

Chivenge, P., Vanlauwe, B., Gentile, R., Wangechi, H., Mugendi, D., van Kessel, C., and Six, J. 2009. Organic and mineral input management to enhance crop productivity in Central Kenya. *Agronomy Journal*, 101(5), 1266–1275. https://doi.org/10.2134/agronj2008.0188x.

Cooke, G. W. 1970. *The Control of Soil Fertility*. Crosby, Lockwood and Son Ltd, London.

Dejene, K., Dereje, A., and Daniel, G. 2010. Synergistic effects of combined application of organic and inorganic fertilizers on the yield and yield components of tef (*Eragrostis tef* (Zucc.) Trotter) under terminal drought at Adiha, Northern Ethiopia. *Journal of the Drylands*, 3(1).

Drechsel, P., Gyiele, L., Kunze, D., and Cofie, O. 2001. Population density, soil nutrient depletion, and economic growth in sub-Saharan Africa. *Ecological Economics*, 38(2), 251–258. https://doi.org/10.1016/S0921-8009(01) 00167-7.

Dunjana, N., Nyamugafata, P., Shumba, A., Nyamangara, J., and Zingore, S. 2012. Effects of cattle manure on selected soil physical properties of smallholder farms on two soils of Murewa, Zimbabwe. *Soil Use and Management*, 28(2), 221–228. https://doi.org/10.1111/sum.2012.28.issue-2.

Eghball, B. 2002. Soil properties as influenced by phosphorus and nitrogen – Based manure and compost applications. *Agronomy Journal*, 94(1), 128–135. https://doi.org/10.2134/agronj2002.0128.

Eghball, B., and Power, J. F. 1999. Phosphorus and nitrogen – Based manure and soil compost application: Corn production and soil phosphorus. *Soil Science Society of America Journal*, 63(4), 895–901. https://doi.org/10.2136/sssaj1999.634895x.

FAO. 2000. *Fertilizers and Their Use*, 4th ed., Food and Agriculture Organization, Rome, Italy.

FAO. 2006a. *Plant Nutrition for Food Security: A Guide for Integrated Nutrient Management*, Food and Agriculture Organization, Rome, Italy.

FAO. 2006b. A guide for integrated nutrient management. Fertilizer and Plant Nutrition Bulletin, no. 16 vol. 16.

FAO. 2010. *Agriculture's Role in Greenhouse Gas Emissions and Capture*. American Society of Agronomy, Crop Science Society of America and Soil Science Society of America, Madison, WI.

FAO and ITPS. 2015. Status of the World's Soil Resources (SWSR)–Main Report. Food and Agriculture Organization of the United Nations and Intergovernmental Technical Panel on Soils, Rome, Italy, 650. Chicago, IL.

Gafar, A. F., Yassin, M., Ibrahim, D., and Yagoob, S. O. 2014. Effect of different (bio organic and inorganic) fertilizers on some yield components of rice (*Oryza sativa* L.). *Universal Journal of Agricultural Research*, 2(2), 67–70.

Giller, K. E., Tittonell, P., Rufino, M. C., Van Wijk, M. T., Zingore, S., Mapfumo, P., Adjei-Nsiah, S., Herrero, M., Chikowo, R., Corbeels, M., Rowe, E.C., Baijukya, F., Mwijage, A., Smith, J., Yeboah, E., van der Burg, W.J., Sanogo, O.M., Misiko, M., and Vanlauwe, B. 2011. Communicating complexity: Integrated assessment of trade-offs concerning soil fertility management within African farming systems to support innovation and development. *Agricultural Systems*, 104(2), 191– 203.

Ginting, D., Kessavalou, A., Eghball, B., and Doran, J.W. 2003. Greenhouse gas emissions and soil indicators four years after manure and compost applications. *Journal of Environmental Quality*, 32(1), 23–32. https://doi.org/10.2134/jeq2003.2300.

Goulding, K., Jarvis, S., and Whitmore, A. 2008. Optimizing nutrient management for farm systems. *Philosophical Transactions of the Royal Society B: Biological Sciences*, 363(1491), 667–680.

Holden, S. T., and Bezabih, M. 2008. Gender and land productivity on rented land in Ethiopia. In *The Emergence of Land Markets in Africa: Impacts on Poverty, Equity and Efficiency*, Routledge, New York, NY (pp. 179–196).

Jahangir, M. M. R., Jahan, I., and Mumu, N. J. 2018. Management of soil resources for sustainable development under a changing climate. *Journal of Environmental Science and Natural Resources*, 11(1–2), 159–170. https://doi.org/10.3329/jesnr.v11i1-2.43383.

John, R., Kapukha, M., Wekesa, A., Heiner, K., and Shames, S. 2014. *Sustainable Agriculture Land Management Practices for Climate Change Mitigation*. EcoAgriculture Partners, Washington, DC.

Lal, R. 2004a. Soil carbon sequestration impacts on global climate change and food security. *Science*, 304(5677), 1623–1627. https://doi.org/10.1126/science.1097396.

Lal, R. 2005a. Forest soils and carbon sequestration. *Forest Ecology and Management*, 220(1–3), 242–258. https://doi.org/10.1016/j.foreco.2005.08.015.

Lal, R. 2005b. Soil carbon sequestration in natural and managed tropical ecosystems. *Journal of Sustainable Forestry*, 2(1–3), 1–30. https://www.sciencedirect.com/science/article/abs/pii/S0378112705004834.

Lal, R. 2005c. Forest soils and carbon sequestration. *Forest Ecology and Management*, 220(1–3), 242–258. https://doi.org/10.1016/j.foreco.2005.08.015.

Lal, R. 2010. Managing soils and ecosystems for mitigating anthropogenic carbon emissions and advancing global food security. *BioScience*, 60(9), 708–721. https://doi.org/10.1525/bio.2010.60.9.8.

Lal, R., 2004b. *Soil Carbon Sequestration to Mitigate Climate Change*. Carbon Management and Sequestration Center, School of Natural Resources. OARDC/FAES, Ohio State University, Columbus, OH.

Mahajan, A., Bhagat, R. M., and Gupta, R. D. 2008. Integrated nutrient management in sustainable rice wheat cropping system for food security in India. *SAARC Journal of Agriculture*, 6(2), 29–32. https://www.researchgate.net/profile/R_Bhagat2/publication/267386083_integrated_nutrient_ management_in_sustainable_rice-wheat_cropping_system_for_food_security_in_india/links/564ec08308aefe619b0ff0d7.pdf

Maro, G. P., Mrema, J. P., Msanya, B. M., and Teri, J. M. 2013. Farmer's perception of soil fertility problems and their attitudes towards integrated soil fertility management for coffee in Northern Tanzania. *Journal of Soil Science and Environmental Management*, 4(5), 93–99. https://doi.org/10.5897/JSSEM.

Milkha, S., and Aulakh, A. 2010. Integrated nutrient management for sustainable crop production, improving crop quality and soil health, and minimizing environmental pollution. Presented at the 2010 19th World Congress of Soil Science, Soil Solutions for a Changing World.

Naresh, R. K., Purushottam, and Singh, S. P. 2013. Effects of integrated plant nutrient management (IPNM) practices on the sustainability of maize-based farming systems in Western Uttar Pradesh. *International Journal of Research in Biomedicine and Biotechnology*, 2(3), 5–10. http://www.urpjournals.com.

Ofosu, A. J., and Leitch, M. 2009. Relative efficacy of organic manures in spring barley (*Hordeum vulgare* L.) production. *Australian Journal of Crop Science*, 3(1), 13–19. https://www.cabdirect.org/cabdirect/abstract/20093037259.

Parry, M.L., Canziani, O., Palutikof, J., Van Der Linden, P. and Hanson, C. (2007) IPCC Climate Change 2007: Impacts, Adaptation and Vulnerability. Contribution of Working Group II to the Fourth Assessment Report of the Intergovernmental Panel on Climate Change. Cambridge University Press, Cambridge, UK, 976.

Philip, J. W., Crawford, J. W., Díaz Álvarez, M. C., and García Moreno, R. 2012. Soil management for sustainable environment. *Applied and Environmental Soil Science Journal*, 196(1), 79–91. https://doi.org/10.1155/2012/850739.

Pinitpaitoon, S., Suwanarit, A., and Bell, R. W. 2011. A framework for determining the efficient combination of organic materials and mineral fertilizer applied in maize cropping. *Field Crops Research*, 124(3), 302–315.

Pretty, J., Toulmin, C., and Williams, S. 2011. Sustainable intensification in African agriculture. *International Journal of Agricultural Sustainability*, 9(1), 5–24. https://doi.org/10.3763/ijas.2010.0583.

Rauniyar, G. P., and Goode, F. M. 1992. Technology adoption on small farms. *World Development*, 20(2), 275–282. https://doi.org/10.1016/0305-750X(92)90105-5

Sainju, U. M., and Good, R. E. 1993. Vertical root distribution in relation to soil properties in New Jersey Pinelands forests. *Plant and Soil*, 150(1), 87–97. https://doi.org/10.1007/BF00779179.

Sainju, U. M., Jabro, J. D., and Stevens, W. B. 2008. Soil carbon dioxide emission and carbon content as affected by irrigation, tillage, cropping system, and nitrogen fertilization. *Journal of Environmental Quality*, 37(1), 98–106. https://doi.org/10.2134/jeq2006.0392.

Sanchez, P. A. 1994. Tropical soil fertility research, towards the second paradigm. In *15th World Congress of Soil Science* (pp. 65–88). Mexican Society of Soil Science, Chapingo, Mexico.

Sanginga, N., Dashiell, K., Diels, J., Vanlauwe, B., Lyasse, O., Carsky, R. J., Tarawali, S., Asafo-Adjei, B., Menkir, A., Schulz, S., Singh, B. B., Chikoye, D., Keatinge, D., and Rodomiro, O. 2003. Sustainable resource management coupled to resilient germplasm to provide new intensive cereal–grain legume–livestock systems in the dry savanna. *Agriculture, Ecosystems and Environment*, 100(2–3), 305–314. https://doi.org/10.1016/S0167-8809(03)00188-9.

Sathish, A., Govinda Gowda, V., Chandrappa, H., and Nagaraja, K. 2011. Long term effect of integrated use of organic and inorganic fertilizers on productivity, soil fertility and uptake of nutrients in rice and maize cropping system. *Inter-national Journal of Science and Nature*, 2(1), 84–88. https://www.researchgate.net/profile/Arun_Ramachandravarapu/publication/316285940_ Biothreats_-_Bacterial_Warfare_Agents/links/58f9e7284585152edece789b/Biothreats-BacterialWarfare-Agents.pdf

Satyanarayana, V., Prasad, P. V., Murthy, V. R. K., and Boote, K. J. 2002. Influence of integrated use of farmyard manure and inorganic fertilizers on yield and yield components of irrigated lowland rice. *Journal of Plant Nutrition*, 25(10), 2081–2090. https://doi.org/10.1081/PLN-120014062.

Scharlemann, J. P., Tanner, E. V., Hiederer, R., and Kapos, V. 2014. Global soil carbon: Understanding and managing the largest terrestrial carbon pool. *Carbon Management*, 5(1), 81–91. https://doi.org/10.4155/cmt.13.77.

Schuijs, M. J., Willart, M. A., Vergote, K., Gras, D., Deswarte, K., Ege, M. J., Madeira, F.B., Beyaert, R., van Loo, G., Bracher, F., and Von Mutius, E. 2015. Farm dust and endotoxin protect against allergy through A20 induction in lung epithelial cells. *Science*, 349(6252), 1106–1110.

Seybold, C. A., Herrick, J. E., and Brejda, J. J. 1999. Soil resilience: A fundamental component of soil quality. *Soil Science*, 164(4), 224–234. https://doi.org/10.1097/00010694-199904000-00002.

Shata, S. M., Mahmoud, A., and Siam, S. 2007. Improving calcareous soil productivity by integrated effect of intercropping and fertilizer. *Research Journal of Agriculture and Biological Sciences*, 3(6), 733–739. http://www.aensiweb.net/AENSIWEB/rjabs/rjabs/ 2007/733-739.pdf.

Simpson, R. J., Richardson, A. E., Nichols, S. N., and Crush, J. R. 2014. Pasture plants and soil fertility management to improve the efficiency of phosphorus fertilizer use in temperate grassland systems. *Crop and Pasture Science*, 65(6), 556–575. https://doi.org/10.1071/CP13395.

Singh, R., and Agarwal, S. K. 2001. Growth and yield of wheat (*Triticum aestivum* L.) as influenced by levels of farmyard manure and nitrogen. *Indian Journal of Agronomy*, 46(3), 462–467. http://www.indianjournals.com/ijor.aspx?target=ijor:ija&volume=46&issue=3&article=015.

Smith, J., Abegaz, A., Matthews, R. B., Subedi, M., Orskov, E. R., Tumwesige, V., and Smith, P. 2014. What is the potential for biogas digesters to improve soil fertility and crop production in Sub-Saharan Africa? *Biomass and Bioenergy*, 70, 58–72. https://doi.org/10.1016/j.biombioe.2014.02.030.

Tabo, R., Bationo, A., Gerard, B., Ndjeunga, J., Marchal, D., Amadou, M., Sogodogo, D., Taonda, J., Hassane, O., Diallo, M., and Koala, S. 2007. Improving cereal productivity and farmers' income using a strategic application of fertilizers in West Africa. In Bationo, A., Waswa, B., Kihara, J., and Kimetu, J. (Eds.), *Advances in Integrated Soil Fertility Management in Sub-Saharan Africa: Challenges and Opportunities* (pp. 201–208). Springer, Dordrecht, NL.

Tulema, B., Aune, J. B., and Breland, T. A. 2007. Availability of organic nutrient sources and their effects on yield and nutrient recovery of tef [*Eragrostis tef* (Zucc.) Trotter] and on soil properties. *Journal of Plant Nutrition and Soil Science*, 170(4), 543–550. https://doi.org/10.1002/(ISSN)1522-2624.

Van Scholl, L., and Nieuwenhuis, R. 2007. *Soil Fertility Management*, 4th ed., Agromisa Foundation, Wageningen, NL.

Vanlauwe, B., Aihou, K., Aman, S., Iwuafor, E. N. O., Tossah, B. K., Diels, J., Sanginga, N., Lyasse, O., Merckx, R., and Deckers, J. 2001a. Maize yield as affected by organic inputs and urea in the West African moist savanna. *Agronomy Journal*, 93(6), 1191–1199. https://doi.org/10.2134/agronj2001.1191.

Vanlauwe, B., Bationo, A., Chianu, J., Giller, K. E., Merckx, R., Mokwunye, U., Ohiokpehai, O., Pypers, P., Tabo, R., Shepherd, K., Smaling, E. M. A., and Woomer, P. L. 2010. Integrated soil fertility management: Operational definition and consequences for implementation and dissemination. *Outlook on Agriculture*, 39(1), 17–24. https://doi.org/10.5367/000000010791169998.

Vanlauwe, B., Diels, J., Aihou, K., Iwuafor, E. N., Lyasse, O., Sanginga, N., and Merckx, R. 2002. Direct interactions between N fertilizer and organic matter. Evidence from trials with 15N-labeled fertilizer. In Vanlauwe, B., Diels, J., Sanginga, N., and Merckx, R., (Eds.), *Integrated Plant Nutrient Management in SubSaharan Africa: From Concepts to Practice* (pp. 256–285). CAB International, Wallingford, UK.

Vanlauwe, B., and Giller, K. E. 2006. Popular myths around soil fertility management in sub-Saharan Africa. *Agriculture, Ecosystems and Environment*, 116(1–2), 34–46. https://doi.org/10.1016/j.agee.2006.03.016.

Vanlauwe, B., Kihara, J., Chivenge, P., Pypers, P., Coe, R., and Six, J. 2011. Agronomic use efficiency of N fertilizer in maize-based systems in sub-Saharan Africa within the context of integrated soil fertility management. *Plant and Soil*, 339(1–2), 35–50. https://doi.org/10. 1007/s11104-010-0462-7.

Watson, C. A., Atkinson, D., Gosling, P., Jackson, L. R., and Rayns, F. W. 2002. Managing soil fertility in organic farming systems. *Soil Use and Management*, 18, 239–247.

Wossen, T., Berger, T., and Di Falco, S. 2015. Social capital, risk preference, and adoption of improved farmland management practices in Ethiopia. *Agricultural Economics*, 46(1), 81–97. https://doi.org/10.1111/agec.2015.46.issue-1.

Zingore, S., Delve, R. J., Nyamangara, J., and Giller, K. E. 2008. Multiple benefits of manure: The key to maintenance of soil fertility and restoration of depleted sandy soils on African smallholder farms, Nut. *Nutrient Cycling in Agroecosystems*, 80(3), 267–282.

9

Monitoring Common Agricultural Cropping Across the US and Canadian Laurentian Great Lakes Basin Watershed Using MODIS-NDVI Data

Ross Lunetta, John G. Lyon, Yang Shao, and Jayantha Ediriwickrema

CONTENTS

Introduction

Corn ethanol production increased rapidly across the midwestern USA from 2005 to 2007 in part due to the Renewable Fuel Standard's Corn Ethanol Mandate (RFA, 2007; Motamed et al., 2016). The environmental implications associated with corn-based ethanol production have received increasing attention (Pimentel, 2003; Pimentel and Patzek, 2005; Zah et al., 2007; Scharlemann and Laurence, 2008). The research of Zah et al. (2007) suggested that corn ethanol may have greater overall environmental cost than using fossil fuels (Scully et al., 2021). Studies have related corn ethanol production

DOI: 10.1201/9781003175018-9

near ethanol plants in the midwestern USA (Motamed et al., 2016) including data showing acreage in corn and overall agriculture not only growing in already-cultivated areas but also expanded into previously uncultivated areas (Motamed et al., 2016).

Water quality, soil erosion, air pollution, crop revenue, biodiversity, and the loss of natural habitats are concerns at both local and regional scales (Hodge, 2002; Huston and Marland, 2003; Pimentel and Patzek, 2005; Searchinger et al., 2008). This is also a crop revenue concern (Katchova and Sant'Anna, 2019) as well as the issues of reenrollment of land into the Conservation Reserve Program (Ifft, Rajagopal and Ryan, 2016; Chen and Khanna, 2018).

Environmental assessments often need site-specific information about crop distributions as statistics and as model inputs (e.g., SWAT, 2007). Researchers are not only interested in the total area of ethanol crop production, but also require data documenting geographic distributions and changes over time to support distributed modeling efforts (Scully et al., 2021). Such information is particularly useful for identifying watersheds subject to potential environmental damages or ecological degradations. Due to the limited availability of National Agricultural Statistics Service (NASS) Crop Data Layer (CDL) products (https://www.nass.usda.gov/Research_and_Science/Cropland/sarsfaqs2, last checked December 10, 2021), researchers often rely on agricultural statistics estimates (i.e., state or county level), developed by the United States Department of Agriculture NASS program (Sheehan et al., 2004; Boryan, 2012). The spatial details of the crop location, extent and distribution, and the pattern of crop change are generally unavailable from the estimated agricultural statistics (Remote Sensing Institute, 1973). Researchers are thus forced to use unrealistic assumptions of crop distributions and crop rotation patterns, which may lead to high uncertainties in modeling predictions of potential environmental impacts.

The mapping of crops using remote sensor data has shown good potential for characterizing the extent, distribution, and condition of croplands (Moran et al. 1997; Frolking et al., 1999; Doraiswamy et al., 2005; Thenkabail et al., 2009). The Moderate Resolution Imaging Spectroradiometer (MODIS) data, which combine moderate spatial resolution (250 m) and a high temporal resolution (1–2 days repeat cycle), were found to be particularly useful to differentiate general cropland versus non-cropland and to categorize individual crop types (Lobell and Asner, 2004; Chang et al., 2007; Wardlow et al., 2007; Wardlow and Egbert, 2008). The phenology-based categorization (or time-series analysis) of MODIS-NDVI (Normalized Difference Vegetation Index) is one of the most used approaches (DeFries and Townshend, 1994; Hansen et al., 2003; Thenkabail et al., 2009). Most previous MODIS-NDVI crop-mapping applications have focused on single-year crop-mapping efforts. MODIS-NDVI datasets have rarely been used to study the crop changes or rotations over multiple years. The potential of multi-year MODIS-NDVI crop mapping has not yet been fully exploited.

The objective was to examine the cropland changes across the US and Canadian Laurentian Great Lakes Basin (GLB) watershed using map products derived from MODIS-NDVI data. The GLB is a region thought to have undergone significant changes in cropping patterns, because the US government implemented substantial subsidies to encourage corn ethanol production during the study period (2005–2007). Research questions of interest included: (1) how did crop acreage distributions (i.e., corn, soybean, and wheat) change through the GLB? (2) Was there a change in crop rotational patterns due to increased corn ethanol demand? (3) If yes, were there any geographic differences associated with variations in crop rotation patterns across the GLB (i.e., US versus Canadian)? The answers to these questions are particularly important for identifying areas or regions with a high potential for environmental degradation.

Two specific research objectives of this chapter were to map annual crop distributions across the GLB for 2005, 2006, and 2007, and compare the 2-year crop change or rotation patterns for 2005–2006 and 2006–2007.

Study Area

The GLB region includes all or part of eight states of the USA and a portion of the Province of Ontario, Canada (Figure 9.1). While the basin is among the most industrialized regions in the world, the southern portions of the GLB are prime areas for corn, soybean, and other types of agricultural crop production (U.S. Environmental Protection Agency, 2008). GLB agricultural production represents 7% and 25% of the total US production and Canadian production, respectively (U.S. Environmental Protection Agency, 2008).

The total agricultural land in the US portion of GLB has decreased slightly from 1970 to 2001 (Erickson, 1995; Wolter et al., 2006). The loss of agricultural land mostly occurred near urban edge areas. For example, southeast Michigan alone lost 13% of agricultural land area from 1990 to 2000, mainly due to the urban expansion (Southeast Michigan Council of Governments, 2003). A majority of Ontario's prime agricultural lands are in the southern part of the province, which have also been subject to extensive urban expansion. Approximately 18% of the prime agricultural land in Ontario has been converted to urban from 1976 to 1996 (Statistics Canada, 1998). However, crop yields associated with major crop types have increased dramatically over the past decades due to technology improvement (Matson et al., 1997).

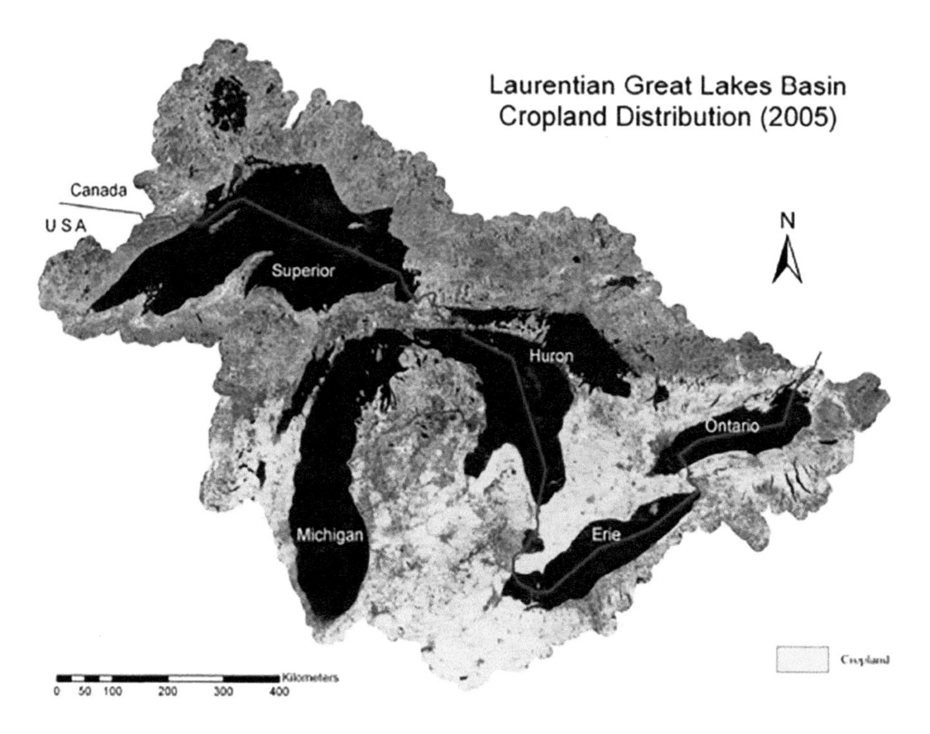

FIGURE 9.1 AND PLATE 9.1
MODIS-NDVI derived cropland extent and distribution across the GLB in 2005.

Urbanization and intensive agricultural production are believed to cause several environmental problems including: (1) sedimentation and excessive nutrient loading (Crosbie and Chow-Fraser, 1999; Midwood and Chow-Fraser 2012); (2) hydrological modifications (Environment Canada and US EPA); and (3) the loss of natural habitats and diminished biodiversity (Detenbeck et al., 1999).

Methods

The MODIS-NDVI data were preprocessed using the method developed by Lunetta et al. (2006). MODIS-NDVI data preprocessing was conducted to provide a filtered (anomalous data removed) and cleaned (excluded data values estimated) uninterrupted data stream to support time-series analysis. MODIS-NDVI 16-day composite grid data (MOD13Q1, version 5.0) in HDF format were acquired between February 2000 and December 2007 from the NASA Earth Observing System data gateway. Details documenting the MODIS-NDVI compositing process and Quality Assessment Science Data

Sets (QASDS) can be found at NASA's MODIS website (MODIS, 1999). NDVI data were subset to the GLB study boundary (10 km buffer), re-projected from a sinusoidal to an Albers Equal-Area Conic projection, using a nearest-neighbor resampling routine, and entered in a 250 m × 250 m grid cell multi-layer image stack. Separate data stacks were developed for both the original NDVI data and QASDS.

The NDVI data stack was first filtered to eliminate anomalous high (hikes) and low (drops) values and then filtered for a second time using the QASDS ratings to remove poor quality data values from the NDVI data stack. Hikes and drops were effectively eliminated by removing data values that suddenly decreased or increased and then immediately returned to near the previous NDVI value. The threshold for the removal of pseudo hikes and drops was optimized at ±0.15% to achieve the best setting (determined qualitatively) to eliminate most all anomalous points, while not inadvertently removing good data points, resulting in a smoother temporal profile. The MODIS QASDS data quality ratings were then applied to retain only those pixels rated as "acceptable" or higher. The filtered data were then transformed into frequency domain using a discrete Fourier transformation and the signal and noise spectrum separated (Roberts et al., 1987; Azzali and Menenti, 2000; Roerink and Menenti, 2000). The removed or potentially corrupted NDVI data points were estimated from the frequency domain signal spectrum using a nonlinear deconvolution approach described by Roberts et al. (1987) to estimate complete "filtered and cleaned" NDVI temporal profiles for each pixel within the GLB.

The filtered and cleaned NDVI temporal profiles provided a "high-quality" dataset that was used to support both the general cropland and crop-specific (e.g., corn, soybean, wheat) mapping. For the cropland versus non-cropland mapping, the 2001 National Land Cover Dataset (NLCD-2001) was obtained to provide training data. We also collected most available Landsat Enhanced Thematic Mapper Plus (ETM+) images covering the time interval of 2000–2002 to construct a seamless image mosaic for the GLB. The Landsat seamless mosaic provided sufficient spatial resolution for a visual interpretation of general cropland and non-cropland classes. This was useful for a validation of cropland mapping results, especially for the Canadian portion of the GLB.

For crop type specific classifications (e.g., corn, soybean, wheat), we obtained all available cropland data layers (CDLs) for 2007 across the US portion of the GLB. The CDL data was primarily developed from Advanced Wide Field Sensor imagery and had a high classification accuracy (>90%) for most major crop types (NASS, 2007). A problem related to the CDL was its limited spatial coverage—primarily focused on the intensive agricultural regions in the midwestern USA. Across the GLB region, only limited spatial coverage of CDLs was available, mainly for the states of Michigan and Ohio. Additionally, we obtained the state and provincial-level agricultural statistics from the NASS and Ontario Ministry of Agriculture, Food, and Rural Affairs (OMAFRA).

General Cropland Mapping

The categorization of general cropland and non-cropland cover types was conducted prior to the crop-specific identifications using training samples derived from the NLCD-2001 (reference data). Specifically, we built a geographic linkage between the NLCD-2001 and the MODIS-NDVI dataset. For each center position of MODIS pixels, we calculated cover proportions for different cover types from the NLCD-2001. Instead of calculating the cover type proportions within the 250 m MODIS-NDVI pixel, we employed a 300 m pixel resolution to reduce the impacts of registration errors. The primary cover type classes considered included water, urban, barren land, forest, shrub/scrub, hay/pasture, cultivated crops, and wetland. Pure MODIS pixels were identified when the 300 m resolution pixel was dominated by one cover type (i.e., >85% homogeneous).

A random sample selection was then conducted to collect MODIS training pixels corresponding to each cover type. The training pixels were further grouped into two general classes of cropland and non-cropland. The total numbers of training pixels were 5,170 and 4,349 for cropland and non-cropland, respectively. It should be noted that the hay/pasture class was included in the cropland group for this study. We did not use the NLCD-2001 to generate a cropland mask, because it only included the US portion of the GLB. It was important to build a general MODIS-NDVI classifier for the entire GLB study area to maintain the categorization consistency across both the USA and Canada.

We used a three-layer MLP (multilayer perceptron) neural network classifier for the categorization of the cropland and non-cropland (Richards and Jia, 1999; Shao et al., 2009). The input layer in the MLP consisted of 13 nodes corresponding to 13 MODIS-NDVI values from Julian days 97 to 289. The MODIS-NDVI values for the remaining dates were discarded due to low information contents (i.e., snow cover in winter). A total number of 15 nodes were used in the hidden layers. The output layers consisted of two nodes indicating the two classes of cropland and non-cropland. The MLP classifier was trained using a threefold stratified cross-validation approach to improve the performance (Duda et al., 2001). The trained network was then employed to classify the MODIS-NDVI image of the entire GLB study area. We assessed the initial cropland map using the Landsat seamless mosaic as reference. Obvious categorization errors were identified, and additional training data points were added iteratively to improve the categorization results. The cropland categorization procedures were conducted for annual MODIS-NDVI datasets corresponding to years 2005, 2006, and 2007.

For the 2005 cropland map, an accuracy assessment was conducted using the Landsat ETM+ seamless mosaic as the primary reference data source. We randomly selected 300 pixels for cropland and non-cropland, respectively. The pixels were visually interpreted from the ETM+ mosaic to assess the accuracy of cropland maps. It should be noted that there was a 3-year time difference between the 2002 ETM+ mosaic and the 2005 MODIS-NDVI cropland map; however, it was the only independent dataset available for the selected study area and time-period.

Crop-Specific Mapping

The identifications of individual crop types were conducted within the cropland mask for each calendar year. Three major crop types were considered including corn, soybean, and wheat. The identification of hay was considered initially, but discarded because of its high phenological variability (i.e., different harvesting time). Hay and the remaining crop types (e.g., sugar beet, potato) were grouped as a mixed class of the "other" crop type. The training pixels were primarily identified using visual interpretation of MODIS-NDVI temporal profiles (Wardlow et al., 2007; Wardlow and Egbert, 2008). For example, the NDVI values for wheat peaked around late-May or early-June, while NDVI values for corn peaked around late-July or early-August. The unique crop phenology provided the basis for differentiating these main crop types using MODIS-NDVI data.

One major challenge associated with the NDVI-based categorization approach was the within-class variability of crop phenology across the large geographic study area. The phenology of a specific crop type (e.g., corn) in northern Michigan may be different from those of Ohio. We employed an ecoregion-stratified categorization approach to improve the categorization performance (Homer et al., 2004). Specifically, we divided the study area into 12 ecoregions using accepted regional ecosystem maps (Omernik, 1987) and conducted independent crop-specific categorizations within each ecoregion. The climate and soil conditions were relatively homogenous within each ecoregion, thus we assumed that crop phenology within each ecoregion was also similar for each of the individual crop types.

The crop-specific categorization was also performed using MLP classifier. The results for the 2007 MODIS-NDVI categorization were assessed using the CDL derived from 2007 Advanced Wide Field Sensor imagery data (56 m) as reference data. We used the same approach described above for the general crop type mapping to assess the performance of the MODIS-NDVI categorization results. The percentages of individual crop types within each 300 m resolution pixel were calculated. Similarly, "pure" pixels were identified if one dominant crop type consisted of >85% homogeneous. All "pure" pixels were used as the reference dataset for the accuracy assessment of MODIS-NDVI crop-specific classification. Overall accuracy and Kappa coefficients were calculated and reported. For classification results corresponding to years 2005 and 2006, there was no reference map available to perform a pixel-wise accuracy assessment. We assumed that categorization accuracies would be similar across study years.

Crop Statistics and Crop Rotation Analysis

The total crop acreages of the GLB were calculated from the crop map products of 2005, 2006, and 2007. The estimated crop acreages from MODIS-NDVI were compared to the state or province-level agricultural statistics obtained from the NASS and OMAFRA. The comparison was conducted for the State of Michigan and the Province of Ontario only, because they are located entirely within the geographic boundaries of the GLB. We did not conduct a county-level comparison, because these data were considered less reliable than state-level values (OMAFRA). More importantly, the county-level agricultural statistics were not complete for all counties or time periods.

The area distributions of three crop types were analyzed at both the GLB and the sub-regional scales. For the sub-regional scale, we simply divided the GLB into US and Canadian sub-regions. The crop distributions in these two sub-regions were compared over the 3-year period. For each of the study years, we also calculated the percentages of corn and soybean areas using 5 km×5 km window. The changes of crop intensity or proportional crop areas from t_1 to t_2 (e.g., 2005–2006) were calculated using the following equation:

$$\Delta P_i = P_{i_2} - P_{i_1} \tag{9.1}$$

where ΔP_i indicates the percent change of crop area in the ith window (5 km×5 km). P_{i_1} and P_{i_2} are the percentage individual crop type area at t_1 and t_2.

At the individual pixel level, the crop rotations were analyzed by stacking 2-year crop maps and conducting a post-categorization comparison. This generated 2-year crop rotation patterns of consecutive years for 2005–2006 and 2006–2007. The frequency of crop rotation patterns (i.e., corn-soybean, soybean-wheat) was calculated within the GLB. This allowed us to identify the most common crop rotation patterns and associated temporal variations.

Results

General Cropland

The overall accuracy for 2005 cropland map was 89% and Kappa coefficient was 0.78. No formal accuracy assessments were conducted for the 2006 and 2007 cropland map products, but we expected similar accuracy levels because the same methods of training data selection and categorization training were employed. The total cropland areas in the GLB were 115,590, 117,973, and 117,352 km^2 for year 2005, 2006, and 2007, respectively. The numbers were relatively stable over the 3-year time-period. Figure 9.1 and Plate 9.1 shows the cropland map from year 2005. The total cropland consisted of approximately 20% of the total land area in the GLB. Over 97% of the croplands were in the southern half of the GLB, especially in the states of Michigan, Ohio, and Wisconsin in the USA and the southern portions of Ontario. Large tracks of land in the northern portion of the GLB remained as forest. Climate and soil quality (i.e., sandy soils) limit large-scale agriculture in the northern GLB region.

An overlay analysis of three cropland maps indicated that about 106,342 km^2 or 92% of cropland pixels remained stable as cropland over the 3-year period; the remaining cropland pixels were labeled as non-cropland for at least one of the years between 2005 and 2007. One possible explanation was that fallow or marginal lands (e.g., conservation reserve program) were involved in conversions between cropland and non-cropland during different calendar years. For example, farmers may leave the land fallow in a certain year. We visually interpreted these "inconsistent" cropland pixels using the ETM+ seamless mosaic and CDL 2007 as reference data. We determined that the majority were spatially scattered, and most were located at the edges of the agricultural land patches, especially at the edges of hay/pasture fields. Therefore, these "inconsistent" croplands might also be associated with classification uncertainties from mixed pixels (Lobell and Asner, 2004). Without available ground truth or reference images across multiple years, it was not possible to provide a more detailed quantitative assessment for these "inconsistent" cropland pixels.

Crop-Specific Mapping

The crop-specific categorization results of 2007 were assessed using CDL 2007 as a reference data source. The CDL 2007 itself was a remote sensing-based crop map product. The accuracies of the CDL were relatively high, especially for major crop types such as corn, soybean, and wheat (>92% in Michigan). For the 2007 crop-specific categorization, the overall accuracy was 84% (Kappa = 0.73). The user's accuracies for corn, soybean, and wheat were 87%, 82%, and 81%, respectively. The producer's accuracies for corn, soybean, and wheat were 85%, 81%, and 83%, respectively. The site-specific accuracy assessment was conducted for year 2007 only, because there were no reference datasets available for years 2005 and 2006.

Table 9.1 shows the comparisons of crop area estimations between the MODIS- NDVI categorizations and statistical estimates obtained from the NASS and OMAFRA. Comparisons were performed to conform with state (Michigan) and province (Ontario) boundaries. Other states were not entirely located within the GLB boundary, thus state-level comparison was not possible. The MODIS-NDVI categorization slightly overestimated corn acreages

TABLE 9.1

Comparisons of MODIS-NDVI Estimated Crop Areas with the NASS (US) and OMAFRA (Canada) Agricultural Statistics for 2005–2007

	2005			2006			2007		
	MODIS	NASS/ OMAFRA	Difference (%)	MODIS	NASS/ OMAFRA	Difference (%)	MODIS	NASS/ OMAFRA	Difference (%)
Michigan									
Corn	9,600	9,105	5.4	8,964	8,903	1.0	10,943	10,724	2.0
Soybean	7,198	8,094	–11.1	7,949	8,094	–1.8	6,280	7,284	–13.8
Wheat	2,245	2,428	–7.5	2,374	2,671	–11.1	2,140	2,226	–3.9
Ontario									
Corn	6,007	6,475	–7.2	6,467	6,385	1.3	8,342	8,498	–1.8
Soybean	8,629	9,409	–8.3	7,831	8,725	–10.2	8,061	9,065	–11.1
Wheat	3,057	3,359	–9.0	3,896	4,162	–6.4	2,243	2,408	–6.9

for Michigan, with the largest difference being observed in 2005 (5.4%). Conversely, the MODIS-NDVI classification underestimated soybean acreages (–1.8% to –13.8%) for both Michigan and Ontario. The estimate discrepancies for corn and soybean were largely due to the confusion between corn and soybean phenological patterns or NDVI profiles (Chang et al., 2007).

The mis-categorization of soybean as corn, or vice versa, was the likely source of confusion. The MODIS-NDVI categorization also underestimated wheat acreages, but the differences between the MODIS estimations and the numbers from the NASS-OMAFRA were generally <10.0%. It should be noted that there were double-cropped fields (wheat harvest in spring and soy harvest in fall). The winter wheat and soybean in NASS-OMAFRA statistics may share the same agricultural field, while the MODIS-NDVI classification produced mutually exclusive classes. This may have resulted in discrepancies in crop area estimations.

Crop Statistics

Table 9.2 shows the crop area distributions for the entire GLB study area over the 3-year period. The total areas of the three crop types were 71,368, 71,900, and 73,586 km^2 for 2005, 2006, and 2007, respectively. These numbers were quite stable over the 3-year study period and corresponded to approximately 62.0% of total croplands in the GLB. The corn acreage decreased by 2.2% from 2005 to 2006, and then increased approximately 21.3% from 2006 to 2007. The soy acreage decreased only slightly (1.0%) from 2005 to 2006, and then further decreased approximately 9.0% from 2006 to 2007. Conversely, wheat acreage increased about 17.9% from 2005 to 2006 and decreased 20.7% from 2006 to 2007.

The change in individual crop acreages altered the general crop area distributions in the GLB (Table 9.2). For example, the area distributions of corn and soybean were almost equal in 2005 (44.1% and 43.8%) and 2006 (42.8% and 43.1%). However, in 2007, the area distribution of corn increased

TABLE 9.2

Crop Areas and Distributions for Corn, Soybean, and Wheat Based on MODIS-NDVI Categorizations throughout the GLB for 2005, 2006, and 2007

	2005		2006		2007	
	Crop Areas % Distribution		Crop Areas % Distribution		Crop Areas % Distribution	
Corn	31,462.4	44.1%	30,765.7	42.8%	37,317.7	50.7%
Soybean	31,283.0	43.8%	30,971.8	43.1%	28,206.9	38.3%
Wheat	8,622.9	12.1%	10,162.9	14.1%	8,062.1	11.0%
Total	71,368.3	100.0%	71,900.4	100.0%	73,586.6	100.0%

to 50.7%, while soybeans decreased to 38.3%. For wheat, the area distribution increased from approximately 12.1% in 2005 to 14.1% in 2006, and then dropped to 11.0% in 2007. We further calculated the crop distributions at the sub-regional level to examine whether there were large spatial variations. Figure 9.2a and b depicts the area distributions of three crop types for the US portion and the Canadian portion of the GLB. For the USA, large increases in corn acreage (19.1%) from 2006 to 2007 resulted in crop acreage decreases for both soybean (–13.1%) and wheat (–7.8%). For the Canadian portion of the GLB (Ontario), there was a 28.2% increase in corn acreage from 2006 to 2007. This coincided with a large decrease in wheat acreage (approximately 40.8%). Figure 9.2c and d shows the area distribution of three crop types for the GLB and the entire USA from national statistics (NASS), respectively. The comparison of Figure 9.2a and d suggests similar cropping change patterns across the US portion of the GLB compared to the US national-level statistics. The large increases in corn acreage from 2006 to 2007, mainly resulted in crop acreage decreases for soybean.

Crop intensities were calculated for both corn and soybeans using 5 km×5 km moving average window. Figure 9.3a–d shows the changes in crop intensity percentage from 2005 to 2006 and 2006 to 2007. From 2005 to 2006, the highest increases of proportional corn areas occurred in the "thumb" area of Michigan, northern Ohio, and south-western Ontario. The proportion of corn subsequently decreased in these areas from 2006 to 2007. It was also evident that corn proportions had increased for most of the other GLB regions from 2006 to 2007. Visual interpretation suggested an inverse relationship between the changes of corn and soybean percentages.

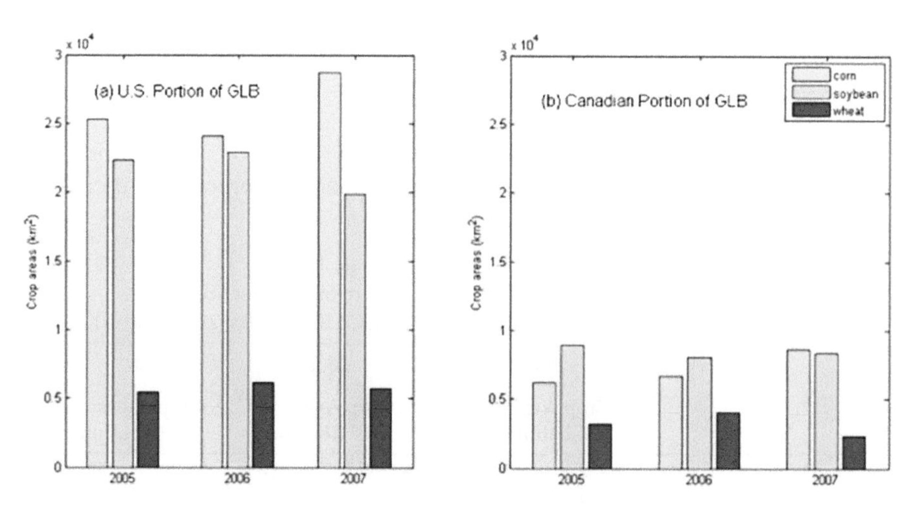

FIGURE 9.2
The total area plantings of corn, soybean, and wheat for the US (a) and Canadian (b) portions of the GLB.

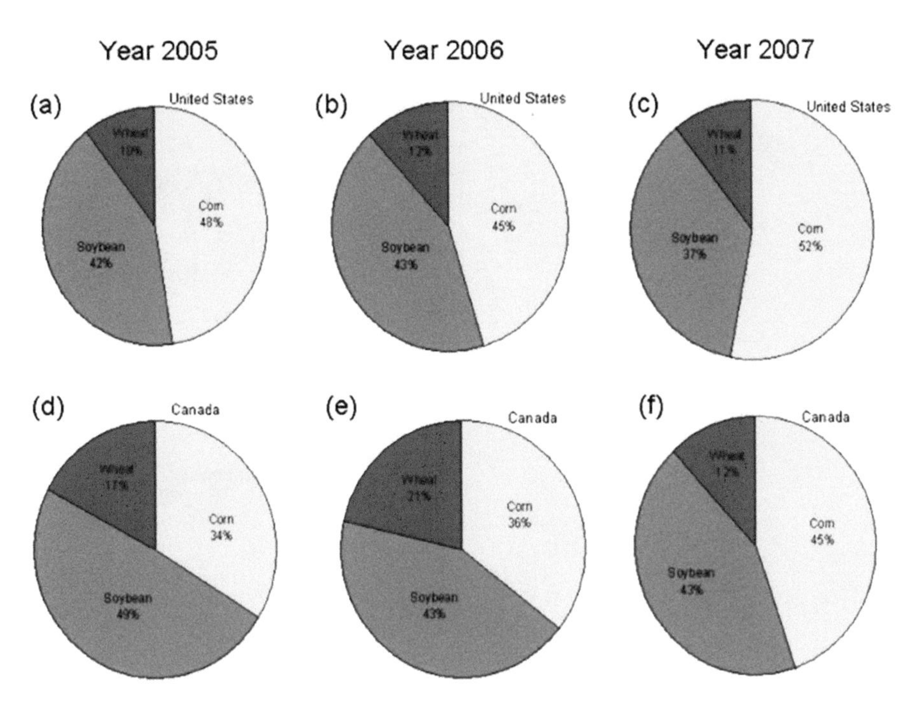

FIGURE 9.3
The percentages of crop cover types for 2005, 2006, and 2007 for the US (a–c) and Canadian (d–f) portions of the GLB.

We further developed cross-plots to compare the changes of cropping intensity for corn and soybeans (Figure 9.4a and b and Plate 9.4). Generally, the changes in corn percentages were negatively correlated with the changes in soybeans for both 2005–2006 (R^2=0.50) and 2006–2007 (R^2=0.53). These results were expected because the corn-soybean rotation was a common agricultural practice in both the USA and Canada. Such rotations take advantage of the nitrogen-fixing capabilities of soybeans to augment the soils. The slope values from a linear regression analysis were –0.76 and –0.65 for 2005–2006 and 2006–2007, respectively. The lower steep slope for the 2006–2007 period suggests that the increases in corn percentages may have also resulted from other crop changes (i.e., other crop to corn) in addition to the typical soybean–corn and soybean–corn-winter wheat rotations (Janovicek et al., 2021).

Crop Rotation Analysis

We aggregated pixels into sixteen categories based on the "from-to" crop rotation or change classes. Tables 9.3a and b summarized the crop

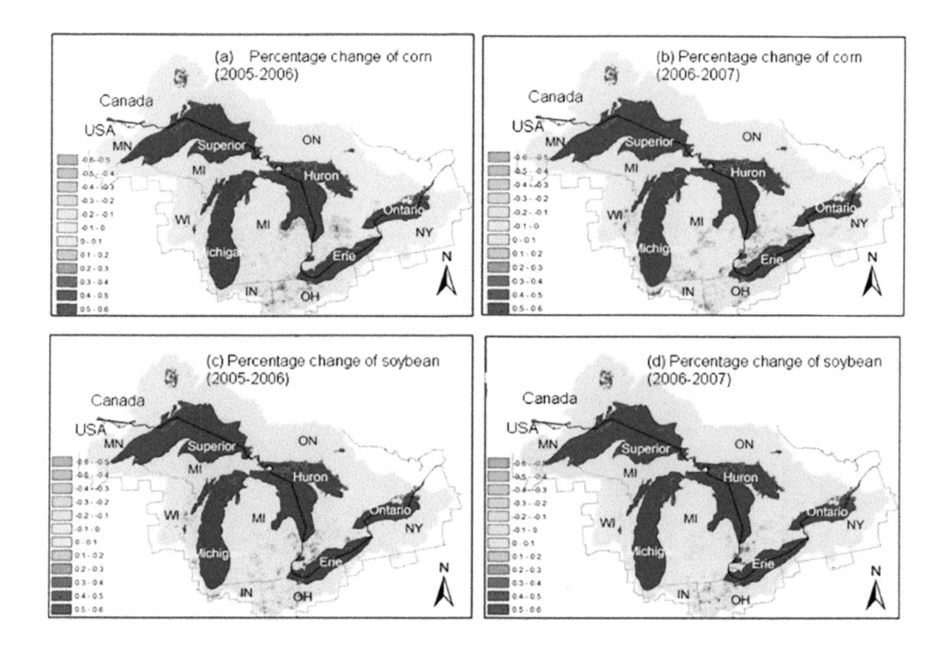

FIGURE 9.4
and Plate 9.4 The percentage crop changes for corn and soybean. Change of corn percentage from 2005 to 2006 (a), (b) change of corn percentage from 2006 to 2007 (b), changes of soybean percentage from 2005 to 2006 (c), and change of soybean percentage from 2006 to 2007 (d).

rotation patterns of the GLB for 2005–2006 and 2006–2007 years, respectively. For 2005–2006, approximately 41.8% of corn fields were converted to soybean fields in 2006 and 24.7% of corn fields in 2005 remained corn fields in 2006. Also 27.8% of corn fields were converted to other crops. For soybean fields in 2005, 40.9% were converted to corn fields. These changes in crop rotation patterns were anticipated because corn-soybean rotation was the most often used crop rotation (Meese et al., 1991; Janovicek et al. 2021). Approximately 22.9% of soybean fields remained as soybean fields in 2006. This percentage was higher than expected because farmers typically did not plant soybean in consecutive years. Categorization confusion (errors) between corn and soybean might be the reason. A post-categorization change detection analysis will generate errors propagating from both individual crop maps (Singh, 1989). An additional source of error could be attributed to mixed pixels containing partial corn and soybean fields. A hard label categorization approach may not be suitable for these mixed pixels (Richards and Jia, 1999). Most wheat fields in 2005 were either converted to soybeans (30.4%) or remained in wheat (28.4%) fields in 2006. For the "other" mixed class type, about 31.4% remained in the same category, while approximately 29.3% were converted to corn and soybean, respectively. The "other" crop type class was a mixed class; thus, it was not possible to identify which specific crop type was involved in the crop rotation.

For 2006–2007 crop rotations, smaller percentages (35.7%) of corn fields were converted to soybean fields compared to 2005–2006 (41.8%). Additionally, 32.6% (2006–2007) versus 24.7% (2005–2006) of corn fields remained in continuous corn production. The percentage of soybean–corn rotation increased to 47.9%, while the percentage of soybean-wheat rotation decreased to 9.2%. The crop rotations of wheat–corn increased to 32.9% as compared to 25.3% for 2005–2006. These results suggested that the large increase in corn acreages in 2007 was responsible for substantial modifications in crop rotation patterns across the GLB from 2006 to 2007.

Discussion

The objectives of this research were to characterize the crop distributions and changes in crop rotations from 2005 to 2007 across the GLB. The MODIS-NDVI 16-day composite product was sufficient to support the development of both general cropland and crop-specific map products at a moderate spatial resolution of 250 m. The spatially explicit crop information was particularly useful for study locations without detailed crop map products (e.g., CDL). For the GLB, the state and provincial-level annual agricultural statistics did not provide spatially explicit data at an adequate temporal frequency to support field level crop inventory or time trajectory analysis.

The MODIS-NDVI crop mapping and analysis provided a cost-effective and timely approach for the regional to sub-regional scale crop change analysis. The total corn acreages from 2005 to 2006 were quite stable. The numbers were also like the total soybean acreages. The corn acreages increased by 21.3% from 2006 to 2007. The increase of corn acreages was directly related to the decrease of soybean acreages (13.1%) in the US portion of the GLB and the decrease of wheat acreages (approximately 40.8%) in the Canadian portion of the GLB. The levels of crop changes greatly exceeded the potential errors attributable to categorization uncertainties. Errors associated with corn and wheat categorizations were generally <10.0% (Table 9.3a and b). The observed crop changes were attributed to cropland changes occurring in the GLB. There were also changes in crop rotation patterns (i.e., increased corn–corn and decreased corn-soybean rotations) for 2006–2007, compared to the 2005–2006 results (Figure 9.5). With current trends of an expanding ethanol industry and high export demand, it is important to monitor crop distributions and crop rotations using timely site-specific approaches, because changes in cropping patterns

TABLE 9.3

Percent Crop Rotations from 2005 to 2006 (a) and 2006 to 2007 (b)

(a)

2006					
	Corn	*Soybean*	*Wheat*	*Other*	*Total*
Corn	24.7%	41.8%	5.6%	27.8%	100%
2005					
Soybean	40.9%	22.9%	13.8%	22.4%	100%
Wheat	25.3%	30.4%	15.9%	28.4%	100%
Other	29.3%	29.4%	9.9%	31.4	100%

(b)

2007					
	Corn	*Soybean*	*Wheat*	*Other*	*Total*
Corn	32.6%	35.7%	5.4%	26.4%	100%
2006					
Soybean	47.9%	24.0%	9.2%	18.8%	100%
Wheat	32.9%	27.4%	9.2%	30.5%	100%
Other	33.9%	26.1%	9.8%	30.1%	100%

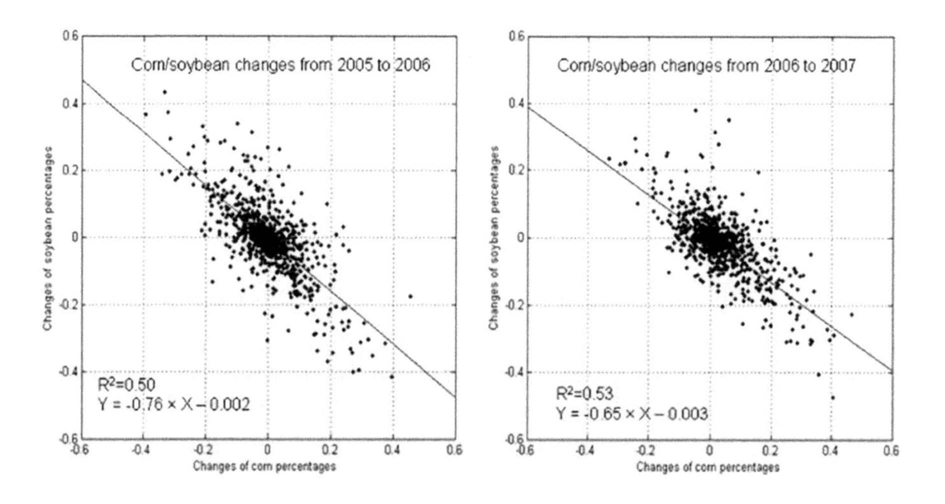

FIGURE 9.5
The relationship between corn change and soybean change from 2005 to 2006 (a) and 2006 to 2007 (b). There were negative correlations between the corn and soybean crop changes. The ordinate is changes of soybean percentages and the abscissa is changes in corn percentages.

may potentially increase non-point pollution to the water, soil erosion, and other environmental costs (Pimentel, 2003). Due to the advantages of the moderate spatial and high temporal resolutions, the MODIS-NDVI-based crop map products are well-suited for environmental assessment applications at watershed or other ecological meaningful units.

It should be noted that the crop changes were only analyzed using a 2-year crop change or rotation period. Although this study identified the most common 2-year crop rotations such as corn-soybean, wheat-soybean, and corn-wheat, it would be difficult to characterize the crop rotational changes that involve three or more sequential years. The greatest challenge would be the cumulative classification errors or error prorogation uncertainty associated with post- categorization comparisons (Singh, 1989). For example, correct identification of 3 years crop rotation would require that three individual pixels (three separate image dates) be accurately labeled. The image misregistration and categorization uncertainty may seriously affect the crop rotation results associated with additional dates (n >2) (Townshend et al., 1992).

Conclusions

MODIS-NDVI datasets were used to develop annual cropland and crop-specific map products for the GLB for growing seasons corresponding to 2005, 2006, and 2007. The crop area distributions and crop rotational changes were characterized by comparing crop map products using two time-steps (2005–2006 and 2006–2007). The area distributions of corn and soybean were almost equal in 2005 and 2006, but there were large increases (21.3%) in corn acreages from 2006 to 2007. Alternatively, soybean and wheat areas decreased approximately 9.0% and 20.7% from 2006 to 2007, respectively. The crop rotation change analyses suggested that the large increase of corn acreages in 2007 introduced changes in crop rotation practices throughout the GLB. These changes resulted in substantial increases of corn–corn, soybean–corn, and wheat–corn crop rotations throughout the GLB. Increases in corn acreages for 2007 were greater in Ontario (28.2%) compared to the USA (19.1%). The increases in corn acreages associated with biofuel mandates could have potential negative impacts on nutrient loadings, pesticide exposures, and sediment-mediated habitat degradation in the USA. The increases in Canadian corn acreages in 2007 were attributed to the higher per bushel corn prices worldwide.

Other studies and authors have developed similar methodologies and found good service. Many of these same sources have found similar trends (Thenkabail et al., 2009; Boryan, 2012; Ifft, Rajagopal and Ryan, 2016; Motamed et al. 2016; Chen and Khanna, 2018; Katchova and Sant'Anna, 2019; Thenkabail et al., 2019).

Acknowledgments

The US Environmental Protection Agency (EPA) funded and partially conducted the research described in this chapter. Although this work was reviewed by EPA and has been approved for publication, it may not necessarily reflect official Agency policy. Mention of any trade names or commercial products does not constitute endorsement or recommendation for use. This research was partially funded by the EPA's Global Earth Observation System of Systems (GEOSS) under the Advanced Monitoring Initiative (AMI) Grant Number 35.

Part of this contribution is based on "Lunetta, R., et al. 2010. Monitoring common agricultural cropping across the US and Canadian Laurentian Great Lakes Basin watershed using MODIS-NDVI data. *International Journal of Applied Earth Observation and Geoinformation*, 12, 81–88" by permission from the publisher Elsevier.

References

Azzali, S., and Menenti, M. 2000. Mapping vegetation-soil-climate complexes in southern Africa using temporal Fourier analysis of NOAA-AVHRR NDVI data. *Int. J. Remote Sens.*, 21(5), 973–996.

Boryan, C., 2012. Remote Sensing of Agriculture NASS' Cropland Data Layer Program. USDA, NASS. https://www.nass.usda.gov/Education_and_Outreach/Reports, _Presentations_and_Conferences/Presentations/Boryan_GMU12_Spring.pdf, last checked November 3, 2021.

Chang, J., Hansen, M.C., Pittman, K, Carroll, M., and DiMiceli, C., 2007. Corn and soybean mapping in the U.S. using MODIS time-series data sets. *Agronomy J.*, 99, 1654–1664.

Chen, X., and Khanna, M. 2018. Effect of corn ethanol production on Conservation Reserve Program acres in the US. *Appl. Energy*, 225, 124–134.

Crosbie, B., and Chow-Fraser, P., 1999. Percentage land use in the watershed determines the water and sediment quality of 22 marshes in the Great Lakes basin. *Can. J. Fish. Aquat. Sci.*, 56, 1781–1791.

DeFries, R. S., and Townshend, J. G. R., 1994. NDVI derived land cover classifications at a global scale. *Int. J. Remote Sens.*, 5, 3567–3586.

Detenbeck, N.E., Galatowitsch, S.M., Atkinson, J. and Ball, H., 1999. Evaluating perturbations and developing restoration strategies for inland wetlands in the Great Lakes Basin. *Wetlands*, 19, 789–820.

Doraiswamy, P.C., Sinclair, T.R., Hollinger, S., Akhmedov, B., Stern, A., and Prueger, J., 2005. Application of MODIS derived parameters for regional yield assessment. *Remote Sens. Environ.*, 97(2), 192–202.

Duda, R.O., Hart, P.E., and Stork, D.G. 2001. *Pattern Classification*. John Wiley & Sons, Inc., NY, p. 654.

Environment Canada and U.S. EPA, 2003. State of the Great Lakes 2003. Environment Canada and the U.S. Environmental Protection Agency, EPA Report No. 905-R-03-004, 114 p.
Erickson D.L., 1995. Rural land-use and land-cover change: implications for local planning in the River Raisin watershed (USA). *Land Use Policy*, 12, 223–236.
Frolking, S., Xiao, X., Zhuang, Y., Salas, W., and Li, C., 1999. Agricultural land-use in China: a comparison of area estimates from ground-based census and satellite-borne remote sensing. *Global Ecol. Biogeogr. Lett.*, 8, 401–416.
Hansen, M.C., DeFries, R.S., Townshend, J.R.G., Carroll, M., Dimiceli, C., and Sohlberg, R.A. 2003. Global percent tree cover at a spatial resolution of 500 m: first results of the MODIS vegetation continuous fields algorithm. *Earth Interact.*, 7, 1–15.
Hodge, C., 2002. Ethanol use in U.S. gasoline should be banned, not expanded. *Oil Gas J.*, 100 (37), 20–30.
Homer, C., Huang, C., Limin, Y., Wylie, B., and Coan, M., 2004. Development of a 2001 national land cover database for the United States. *Photogramm. Eng. Remote Sens.*, 70(7), 829–840.
Huston M.A, and Marland, G., 2003. Carbon management and biodiversity. *J. Environ. Manage.*, 67, 77–86.
Ifft, J., Rajagopal, D., and Ryan, W., 2016. The effect of the ethanol mandate on the Conservation Reserve Program. In 2016 Annual Meeting, July 31-August 2, Boston, Massachusetts, no. 236178. Agricultural and Applied Economics Association, 2016.
Janovicek, K., Hooker, D., Weersink, A., Vyn, R., and Deen, B. 2021. Corn and soybean yields and returns are greater in rotations with wheat. *Agronomy J.*, 113(2): 1691–1711. https://doi.org/10.1002/agj2.20605
Katchova, A., and Sant'Anna. A., 2019. Impact of ethanol plant location on corn revenues for U.S. Farmers, *Sustainability*, 11(22), 1–13.
Lobell, D.B., and Asner, G.P., 2004. Cropland distributions from temporal unmixing of MODIS data. *Remote Sens. Environ.*, 93, 412–422.
Lunetta, R.S., Knight, J.F., Ediriwickrema, J., Lyon, J., and Worthy, L.D., 2006. Land-cover change detection using multi-temporal MODIS NDVI data. *Remote Sens. Environ.*, 105, 142–154.
Matson, P.A., Parton, W.J., Power, A.G., and Swift, M.J., 1997. Agricultural intensification and ecosystem properties. *Science*, 277, 504–509.
Meese, B.G., Carter, P., Oplinger, E., and Pendleton, J. 1991. Corn/soybean rotation effect as influenced by tillage, nitrogen, and hybrid/cultivar. *J. Prod. Agric.*, 4, 74–80.
Midwood, J., and Chow-Fraser, P. 2012. Changes in aquatic vegetation and fish communities following five years of sustained low water levels in coastal marshes of eastern Georgian Bay, Lake Huron. *Global Change Biol.*, 18(1):93–105
MODIS, 1999. MODIS Vegetation Index (MOD 13): Algorithm Theoretical Basis Document, Version 3. http://modis.gsfc.nasa.gov/data/atbd/atbd_mod13.pdf, last checked November 3, 2021.
Moran, M. S., Inoue, Y., and Barnes, E. M., 1997. Opportunities and limitations for image-based remote sensing in precision crop management. *Remote Sens. Environ.*, 61, 319–346.
Motamed, M., McPhail, L., and Ryan Williams, R. 2016. Corn area response to local ethanol markets in the United States: A grid cell level analysis. *Am. J. Agric. Econ.*, 98, 726–743.

NASS (National Agriculture Statistical Service), 2007. USDA - National Agricultural Statistics Service Homepage, last checked November 3, 2021.

Omernik, J.M., 1987. Ecoregions of the conterminous United States. *Ann. Am. Assoc. Geogr.*, 77(1), 118–125.

Pimentel, D., 2003. Ethanol fuels: energy balance, economics, and environmental impacts are negative. *Nat. Resource Res.*, 12, 127–134.

Pimentel, D., and Patzek, T.W., 2005. Ethanol production using corn, switchgrass, and wood; biodiesel production using soybean and sunflower. *Nat. Resource Res.*, 14, 67–76.

Remote Sensing Institute, 1973. The improvement of remote sensing technology and its application to corn yield estimates. Final report to SRS, USDA, South Dakota State University, Brookings, SD, 100 p. https://www.nass.usda.gov/Education_and_Outreach/Reports, _Presentations_and_Conferences/Yield_Reports/The%20Improvement%20of%20Remote%20Sensing%20Technology%20and%20Its%20Spplication%20to%20Corn%20Yield%20Estimates%20(Pages%201-100).pdf, last checked November 3, 2021.

RFA (Renewable Fuels Association), 2007. Ethanol industry statistics. https://ethanolrfa.org/resources/annual-industry-outlook, last checked November 3, 2021.

Richards, J. A., and Jia, X., 1999. *Remote Sensing Digital Image Analysis*. Springer, Dordrecht, NL, p. 363.

Roberts, H., Lehar, J., and Dreher, J.W. 1987. Time series analysis with CLEAN. I. Derivation of a spectrum. *Astron. J.*, 93(4), 968–989.

Roerink, G.J., and Menenti, M. 2000. Reconstructing cloudfree NDVI composites using Fourier analysis of time series. *Int. J. Remote Sens.*, 21(9), 1911–1917.

Scharlemann, J.P.W., and Laurence, W.F., 2008. How green are biofuels? *Science*, 319, 43–44.

Scully, M., Norris, G., Alarcon Falconi, T., and MacIntosh, D. 2021. Carbon intensity of corn ethanol in the United States: state of the science. *Environ. Res. Lett.*, 16(4), 043001.

Searchinger, T., Heimlich, R., Houghton, R.A., Dong, F., Elobeid, A., Fabiosa, J., Tokgoz, S., Hays, D., and Yu, T-H., 2008. Use of U.S. croplands for biofuels increases greenhouse gases through emissions from land-use change. *Science*, 319, 1238–1240.

Shao, Y., Lunetta, R.S., Ediriwickrema, J., and Iiames, J., 2010. Mapping cropland and major crop types across the Great Lakes Basin using MODIS-NDVI data. *Photogramm. Eng. Remote Sens.*, 76, 73–84.

Sheehan, J., Aden, A., Paustian, K., Killian, K., Brenner, J., Walsh, M., and Nelson, R., 2004. Energy and environmental aspects of using corn stover for fuel ethanol. *J. Ind. Ecol.*, 7, 117–146.

Singh, A. 1989. Digital change detection techniques using remotely-sensed data. *Int. J. Remote Sens.*, 10, 989–1003.

Southeast Michigan Council of Governments, 2003. *Land Use Change in Southeast Michigan: Causes and Consequences*. SEMCOG, Detroit, MI.

Statistics Canada, 1998. National Population Health Survey. University of Toronto, Data Library. https://onesearch.library.utoronto.ca/content/map-data-library, last checked November 3, 2021.

SWAT (Soil and Water Assessment Tool), 2007. http://www.brc.tamus.edu/swat/, last checked November 3, 2021.

Thenkabail, P., J.G. Lyon, H. Turral, and H. Biradar, 2009. *Remote Sensing of Global Croplands for Food Security*. Taylor & Francis Series in Remote Sensing Applications, CRC Press, Boca Raton, FL, p. 510.

Townshend, J. R. G., Justice, C. O., Gurney, C., and McManus, J. 1992. The impact of misregistration on change detection. *IEEE Trans. Geosci. Remote Sens.*, 30(5), 1054–1060.

U.S. Environmental Protection Agency, 2008. Explore our Multimedia Page. http://www.epa.gov, last checked December 11, 2021.

Wardlow, B.D., and Egbert, S.L., 2008. Large-area crop mapping using time-series MODIS 250 m NDVI data: an assessment for the U.S. Central Great Plains. *Remote Sens. Environ.*, 112(3), 1096–1116.

Wardlow, B.D., Egbert, S.L, and Kastens, J.H., 2007. Analysis of time-series MODIS 250 m vegetation index data for crop classification in the U.S. Central Great Plains. *Remote Sens. Environ.*, 108(3), 290–310.

Wolter, P.T., Johnston, C.A., and Niemi, G.J., 2006. Land use land cover change in the U.S. Great Lakes basin 1992 to 2001. *J. Great Lakes Res.*, 32, 607–628.

Zah, R., Hischer, R., Gauch, M., Lehman, M., Boni, H., and Wagner, P. 2007. *Okobilanz von Engergieprodukten: Okologische Bewertung von Biotreibstoffen*. Bundesamt für Energie, St. Gallen, Switzerland.

10

SWAT Modeling of Sediment Yields for Selected Watersheds in the Laurentian Great Lakes Basin

Yang Shao, Ross Lunetta, Alexander J. Macpherson, Junyan Luo, and Guo Chen

CONTENTS

Introduction

The US Midwest has experienced significant changes in agricultural cropping patterns (i.e., area and rotation pattern changes) since 2005. Ongoing agricultural land use change is likely to be partly due to rising corn prices and subsidies implemented by the US government to encourage corn ethanol production. The US Department of Agriculture's (USDA) National Agricultural Statistics Service (NASS) reported that corn acreage in 2007 reached the highest level (37.9 million ha) since 1944. The expanding corn acreage is often related to the decrease of other agriculture crops (i.e.,

DOI: 10.1201/9781003175018-10

soybean and winter wheat) and pasture land (Westcott, 2007; Keeney and Hertel, 2009). Remote sensing-based crop rotation study indicated that traditional crop rotation (i.e., corn-soybean) is being replaced by continuous corn plantings (Stern et al., 2008; Lunetta et al., 2010; Secchi et al., 2011) across the Great Lakes Basin (GLB). Shifts toward more intensive corn production may cause several negative environmental consequences with respect to water quality, soil fertility, biodiversity, and overall ecosystem sustainability (Pimentel and Patzek, 2005; Searchinger et al., 2008). For instance, Donner and Kucharik (2008) have raised concerns of corn-based ethanol production with respect to the goal of reducing nitrogen export by the Mississippi River.

Many remote sensing cropland mapping efforts have produced crop-type distributions using a variety of remote sensor imagery, mapping schemes, and image classification algorithms. For example, NASS generated the cropland data layer (CDL) products using Advanced Wide Field Sensor imagery (Johnson, 2008). The Moderate Resolution Imaging Spectroradiometer (MODIS) data also show high potential for mapping individual crop types (Chang et al., 2007; Wardlow et al., 2007; Shao et al., 2010). Most of the above remote sensing efforts focused on characterizing cropland distributions and monitoring change. The impacts of agricultural change on water quality, soil erosion, and biodiversity are still poorly understood.

The integration of land cover changes and watershed modeling provides a useful framework to assess the environmental consequences of agricultural land use change. Tong and Chen (2002) quantified the relative impacts of land uses on surface water quality. They identified agricultural and impervious urban lands as the major source areas for nitrogen and phosphorus loadings to a watershed within the Little Miami River Basin, Ohio. For the Little Eagle Creek watershed in Indiana, Bhaduri et al. (2000) reported that about 80% of the annual runoff increase was due to the expanding impervious surface. Fohrer et al. (2001) calibrated and validated SWAT (Soil and Water Assessment Tool) models for four watersheds and found that the impact of land cover change on the annual water balance was small due to compensating effects in complex catchments. Tang et al. (2005) integrated a land use change model and a web-based environmental impact model to assess the changes in runoff and nutrient loadings due to urbanization. Distributed watershed models are increasingly used to assess the impacts of land use change on hydrologic responses (Miller et al., 2002; Naef et al., 2002), sediment loadings (Allan et al., 1997; Tong and Chen, 2002), nutrient loadings (Allan et. al., 1997; Weller et al., 2003; Lunetta et al., 2005), and in-stream habitat structure (Allan et al., 1997).

Study Objectives

The overall goal of this research was to examine how agricultural land use change affects sediment yields for selected watersheds in the Great Lakes Basin (GLB). The GLB is a region that has undergone significant changes in cropping patterns since 2005 (Lunetta et al., 2010). The currently expanding corn acreage in the GLB might increase sediment and nutrient loadings to the GLB streams, affecting the sensitive GLB ecosystem (GLC, 2007). We are interested in how the spatial distributions of corn planting affect sediment yields. Such information will enable conservation organizations and government agencies to better understand the consequences of environmental and energy policies in agricultural and forested landscapes. The specific research procedures included: (1) implement SWAT models for the selected watersheds to estimate baseline sediment yields using current land use, and (2) predict sediment yields for simulated future agricultural land use conditions.

Study Area

The GLB covers an area of 764,568 km^2 and includes both the USA and Canada. The US portion of the GLB includes all or part of eight states and the Canadian portion includes part of the Province of Ontario. The GLB is one of the most industrialized regions in the world. For the last 30 years, rapid land use change, especially urban growth, and residential sprawl, has raised many issues and concerns with respect to the sustainability of the GLB's ecosystems (USEPA, 2008). The US EPA reported a decrease of 9.5% in agricultural land within the US portion of GLB from 1981 to 1992 (USEPA, 1997). Most of these agricultural lands were converted to urban use. During 1992–2001, there was an additional 2.3% decrease for both agricultural and forested lands, also substantially attributed to urban development (Wolter et al., 2006). In the Canadian portion of the GLB, Statistics Canada (1998) estimated that 18% of agricultural land was converted to urban from 1976 to 1996. Impacts of land cover conversions on the Basin's water quality, biodiversity, and ecosystem sustainability have been a focus of early attention (Crosbie and Chow-Fraser, 1999; Detenbeck et al., 1999; EC and USEPA, 2003).

Under current national and state energy policy, farmers in the region are altering agricultural land use strategies. For example, the corn acreage in the GLB increased approximately 21% from 2006 to 2007, mainly at the cost of soybean and winter wheat acreage (Lunetta et al., 2010). Corn-related crop rotation change (i.e., continuous corn plantings) was also evident (Lunetta et al., 2010). Recent changes to agricultural practices in the GLB are complicating the study of non-point source (NPS) pollution.

Watershed Assessment and the SWAT Model

A variety of hydrologic and water quality models have been used to assess the impacts of land use/land cover changes (Bhaduri et al., 2000; Fohrer et al., 2001; Weller et al., 2003; Tang et al., 2005). For example, Weller et al. (2003) developed an empirical linear model to predict water quality using the proportion of cropland and developed land as independent variables. Bhaduri et al. (2000) integrated GIS with an NPS pollution model to assess the long-term runoff and NPS pollution. Tang et al. (2005) implemented the Long-Term Hydrologic Impact Assessment model (Harbor, 1994), to estimate the impacts of land use changes on surface runoff and NPS pollution. Empirical water quality models have advantages in data preparation as the input data are often readily available and the model can be routinely used for operational applications.

Recently, distributed watershed models have been increasingly used to assess hydrologic responses to different land cover changes. Additionally, process-based watershed models can be very useful for improving the understanding of interactions between land use change, water balance, and water quality issues. The SWAT model is a physically based, continuous time-step model (Arnold et al., 1998; Neitsch et al., 2002; Ward et al., 2015). The model was developed by USDA Agricultural Research Service (ARS) to assess the impact of agricultural management practices on water balance, sediment, and nutrient loadings for non-gauged watersheds. The SWAT model has been widely used in both US and international sites (Ward et al., 2015). Borah and Bera (2004) provided an overview of SWAT applications for 17 case studies. Most SWAT models were calibrated and validated at monthly intervals. Good results were achieved for both small (Warner Creek, 3.46 km^2) and large watersheds (Upper Mississippi River Basin, 491,700 km^2). The daily estimations from SWAT are generally considered less accurate compared to monthly estimations (Borah and Bera, 2004). A thorough review of SWAT model is also provided by Gassman et al. (2007).

One of the main drawbacks of the application of SWAT is its significant data requirements. Primary SWAT input data include a land cover, digital elevation model (DEM), soil map, daily precipitation and temperature, and detailed agricultural management information (land use). The SWAT model's calibration and validation procedures require additional datasets such as stream flow, sediment, and nutrient loadings. Due to limited data availability, thorough model calibration and validation were not possible for many applications (Stonefelt et al., 2000). However, recent advances in remote sensing and GIS have resulted in improved data availability, and continuous software development (i.e., ArcSWAT) has made the SWAT toolbox more user-friendly. As a result, it is expected that SWAT will be widely used for future watershed assessments, particularly those linking land cover and water quality.

Methods

Watershed Selection

The US portion of the GLB consists of 157 USGS 8-digit hydrologic units or watersheds. Only 15 of the 157 watersheds have relatively large portions of agricultural land (i.e., >15%). Within these 15 watersheds, we selected four for SWAT model assessment: the St. Joseph River watershed in the Lower Peninsula of MI and northwestern portion of IN, St. Mary's watershed near the OH-IN border, the Peshtigo River Watershed in Northern WI, and the Cattaraugus Creek Watershed in Western NY (Table 10.1). These four watersheds are in four different ecoregions (Eastern Corn Belt Plains, Southern Michigan/Northern Indiana Drift Plains, Northern Lakes and Forests, and Northern Allegheny Plateau, respectively). Each ecoregion has different climate, soil, and land use conditions (Figure 10.1). The area of the watersheds ranged from 1,430 to 12,132 km^2. The percentage of agricultural land ranged from 14% to 84%. The large variation of agricultural proportions allows us to evaluate the impacts of agricultural land use change on sediment yields. It should be noted that the St. Joseph River watershed is substantially larger than the other three. To better compare across watersheds, we selected a subset of the St. Joseph River watershed (Dry Run Creek) for the SWAT implementation.

TABLE 10.1

The Four Selected Watersheds Located within the GLB and Associated Land Use Characteristics (Ag=Agriculture, HUC=8-Digit)

Name	HUC Ids.	State	Area (km^2)	Ag (km^2)	%Ag
St. Joseph	4050001	MI, IN	12,131.76	7,097.36	68
St. Mary's	4100004	IN, OH	2,108.81	1,649.50	84
Peshtigo	4030105	WI	3,031.08	450.35	14
Cattaraugus	4120102	NY	1,429.59	508.96	35

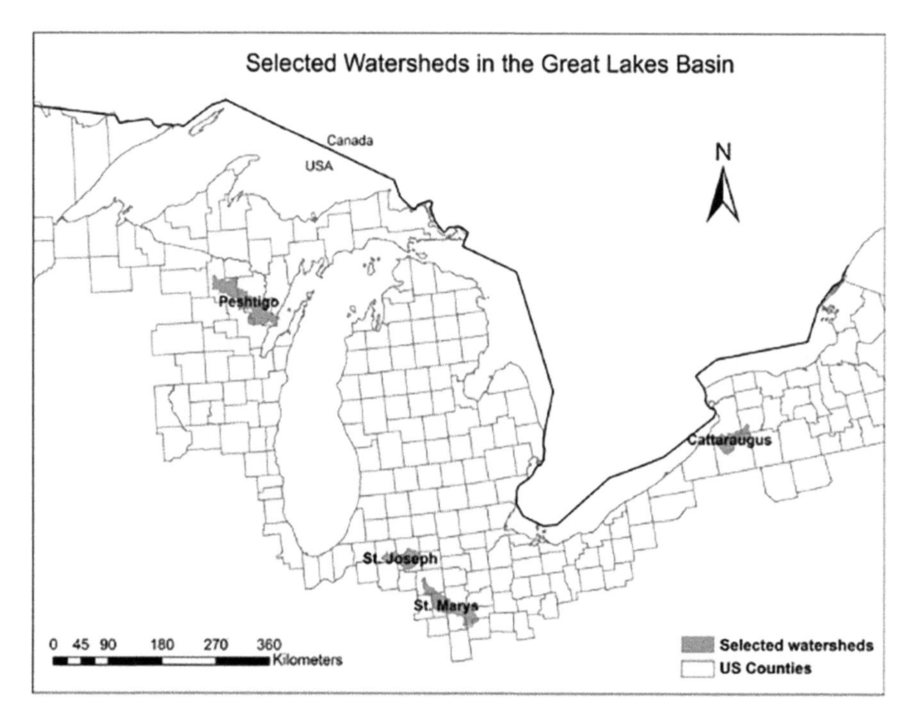

FIGURE 10.1
The four selected watersheds in the GLB.

SWAT Input Data

The 30 m DEM was obtained from the USGS Seamless Data Distribution System. The Soil Survey Geographic data were obtained from the US Natural Resources Conservation Service. The climate data, including daily precipitation and daily temperature, were obtained from the USDA-ARS (Agricultural Research Service). The USDA-ARS climate data were developed using data from the National Oceanic and Atmospheric Administration. The USDA-ARS climate data were already processed using the standard SWAT model input formats, so they can be readily incorporated for SWAT application.

SWAT model is designed to parameterize and analyze a wide range of land use and crop management information (i.e., crop rotation, planting data, tillage, and fertilizer application). However, it is often difficult to obtain detailed agricultural land use and crop management information, especially when multiple watersheds or large study areas are involved (Borah and Bera, 2004).

We used the 2001 National Land Cover Dataset (NLCD) as the primary land use and land cover data. Within the 2001 NLCD row crop areas, the USDA-NASS CDL was used for the supplement crop rotation construction. For the St. Joseph River watershed and a large portion of the St. Mary's

TABLE 10.2

The Primary Crop Rotation and Tillage Practices Applied for Selected Watersheds (c=Corn, s=Soybean, h=Hay)

		Crop Rotation	Tillage
		Corn	Soybean
St. Joseph	c-s, s-c, c-c	Conventional and conservation	Conservation
St. Mary's	c-s, s-c, c-c	Conventional	Conservation
Peshtigo	c-c-c, c-c-h, c-h-c, h-c-c, c-c-s	Conservation	Conservation
Cattaraugus	c-c-c, c-c-h, c-h-c, h-c-c, c-c-s	Conventional	Conventional

watershed, annual CDL data are available since 2000. Three dominant crop rotation patterns (i.e., corn-soybean, soybean-corn, continuous corn) were identified (Table 10.2). For the Peshtigo watershed, there is yearly CDL data since 2003. Continuous corn and corn-alfalfa rotations were the most common crop rotation practices. For the Cattaraugus Creek, there was no corresponding CDL coverage. A combination of continuous corn and corn-alfalfa rotation was assumed for the study watershed. These baseline crop rotation patterns were implemented through permutations within SWAT model (Gassman et al., 2003).

Tillage practice data were obtained from the Conservation Technology Information Center (CTIC). Percentages of tillage practices for corn, soybean, and other major crop types are available through CTIC website. For this study, the 2004 tillage practice information (i.e., no-till, conventional tillage) for corn and soybean were derived for all counties that intersect the watersheds. Within each county, the dominant tillage practice for corn and soybeans were identified and incorporated into the SWAT model. Table 10.2 shows the primary crop rotations and tillage practices used for different watersheds.

SWAT Calibration and Validation

We used ArcSWAT to model water and sediment yields (Winchell et al., 2007). The USGS National Hydrology Dataset (1:100,000 scale) was directly overlaid on the DEM in the watershed delineation procedure to ensure that the stream locations were correctly identified. A threshold value (1,000 ha) was used to define the minimum drainage area required to form a stream branch. The outlet for each watershed was manually selected. The watershed delineation generated a range of GIS layers (i.e., sub-basin, reach) and detailed reports with respect to the topographic aspect of the watershed. For the hydrologic response unit (HRU) definition, we used threshold values of 5%, 10%, and 5% for land cover, soil, and slope class percentages, respectively. These threshold values were used to remove minor land use and soil types, so a simplified HRU definition could be achieved (FitzHugh and Mackay,

2000). Daily precipitation, daily minimum temperature, and daily maximum temperature were derived from the USDA-ARS climate data for the period from January 1999 to December 2008. For the selected four watersheds, all available weather stations within a watershed or in proximity were used as the input. This allowed a better spatial representation for precipitation and temperature data. We used the SWAT default dataset for wind, solar radiation, and relative humidity variables.

We focused on water balance and stream flow calibration for the SWAT model since hydrology is the driving force regulating sediment yields. For all four selected GLB watersheds, the USGS stream flow observation historical data records were available. We obtained data from January 1999 to December 2008 at gauge stations for Elkhart River at Goshen, IN (4100500), St. Mary's River near Fort Wayne, IN (4182000), Peshtigo River at Peshtigo, WI (4069500), and Cattaraugus Creek at Gowanda, NY (4213500). For these four selected watersheds, calendar year 1999 was used for the SWAT model "warm-up" period, stream flow calibration was conducted from years 2000 to 2005, and model validations were conducted for 2006 to 2008.

We followed the recommended SWAT Manual for stream flow calibration (Neitsch et al., 2002). The first calibration step was to compare the average annual observed stream flow and SWAT simulated results. In this procedure, it is often required to estimate the fractions of baseflow and surface runoff from observed data. The baseflow filter program was used to estimate the ratio of surface runoff to baseflow (Arnold and Allen, 1999). The SWAT outputs were required to match the surface runoff and baseflow derived from the observation data.

The curve number (CN2) parameter was used to increase or decrease the SWAT estimated surface runoff. The soil evaporation compensation factor was also adjusted if the curve number alone did not generate good estimates. The parameters for the baseflow calibration included Alpha_BF (baseflow recession constant), GW_Revap (ground water "revap" coefficient), Revapmn (water level in shallow aquifer), and Rchrg_Dp (aquifer percolation coefficient). The annual stream flow calibration procedures were repeated until satisfactory results were achieved (i.e., within 5% difference). We assumed that the monthly variations would be acceptable if the annual stream flow calibration was successful. However, the initial comparison for the monthly stream flow data showed relatively large scattering. Additional SWAT parameters such as SFTMP (snowfall temperature), SMTMP (Snow melt base temperature), SURLAG (surface runoff lag coefficient), N (Manning's coefficient), and TIMP (snow pack temperature lag factor) were also adjusted to improve the SWAT model performance. We targeted an R^2 value of 0.7 as the threshold value for monthly stream flow calibration.

The SWAT predicted stream flow was assessed for the validation period (2006–2008) for monthly intervals. The two most used quantitative measures, the linear regression coefficient of determination (R^2) and the Nash and Sutcliffe model efficiency coefficient (E), were calculated. Detailed calibration and validation for the sediment yields were not feasible due to the limited availability of observation data. For overall comparison purposes, we reviewed the literature for similar watershed studies in the GLB. The SWAT sediment yields were calibrated based on the annual average values.

Future Land Cover Scenarios

The SWAT calibration models were used to assess the sediment yields for two simulated future land cover scenarios. The first scenario was to convert all "other" agricultural row crop types (i.e., sorghum) to corn fields and switch the current/baseline crop rotation into continuous corn. The second scenario was to further expand the corn planting to hay/pasture field. The tillage practices remained to be the same as those of the baseline condition. Although these assumed agricultural scenarios are likely unrealistic, our intention was to assess the boundary conditions under these extreme scenarios. We replaced the current or baseline land use data with the future land use data, while other SWAT model inputs and parameters were held at the baseline condition. The sediment yields for the future land use scenarios were then compared to the baseline sediment yields.

Results and Discussion

SWAT Stream Flows

Independent SWAT models were developed for each of the watersheds. Simple land cover distribution analysis using the NLCD showed that the dominant cover type for the St. Joseph River and St. Mary's watersheds was agricultural row crops (>50%). The Peshtigo River and the Cattaraugus Creek Watersheds were forest-dominated (>50%), although row crops accounted for approximately 15%–20% for both watersheds. For all four watersheds, urban development occupies relatively small portion of the total area (4%–8%). The SWAT watershed delineation procedure created total numbers of 57, 101, 147, and 74 sub-basins for the St. Joseph River, St. Mary's, the Peshtigo River, and the Cattaraugus Creek Watersheds, respectively. The combination of land cover and soil types further delineated 1,094, 1,557, 2,000, and 2,766 HRUs for these four watersheds, respectively. The HRUs are the basic processing units in the SWAT model.

The baseflow filter program estimated that baseflow contributed about 40%–60% of total stream flow for the four watersheds. For example, the baseflow of St. Mary's watershed contributed about 40% of total flow. For the calibration period (2000–2005), the average observed annual baseflow and surface runoff were 178 and 267 mm/yr, respectively. It should be noted that these values were averaged over the entire watershed. Using the default parameters, the SWAT model predicted 102 and 303 mm/yr for baseflow and surface runoff, respectively. It appeared that the SWAT overestimated surface runoff values while underestimating the baseflow. We slightly reduced (–0.5%) the curve number (CN2) to decrease the surface runoff. We also adjusted GW_REVAP (0.02) and REVAPMN (10) to increase the base flow. The SWAT predicted new values of 155 and 244 mm/yr for base flow and surface runoff, respectively. The SWAT predicted monthly stream flows were then compared with the observed values resulting in an R^2=0.68. To achieve the targeted threshold value (R^2=0.70), we tested adjusting several SWAT parameters. Literature review of SWAT applications and EPA internal reports (Ambrosio et al., 2007) were particularly useful for identifying the most used parameters and corresponding values. For the St. Mary's watershed, SFTMP (1.1), SMTMP (3.5), and SURLAG (0.5) appeared to have the highest impacts on the SWAT model performance. The calibration of these parameters was conducted in an iterative manner until acceptable calibration results were achieved. For example, the default SURLAG value is 4 days. We adjusted the SURLAG value in the range of 0.5–6. We found that the time of concentration of 0.5-day produced the best results. Manning's coefficient for the tributary channel was adjusted from 0.014 to 0.05 (Neitsch et al., 2002). The R^2 values increased to 0.83 after the model calibration. This value was much higher than the recommended threshold (R^2=0.5) by Gassman (2008) and Nair (2010).

The same calibration procedures were conducted for the other three watersheds. Table 10.3 shows the common parameters adjusted in the calibration. All parameters required iterative testing to achieve the satisfactory calibration results. Overall, the calibration of stream flow achieved R^2 values of 0.71, 0.83, 0.69, and 0.67 for the St. Joseph River, St. Mary's, the Peshtigo River, and the Cattaraugus Creek Watersheds, respectively. The corresponding E

TABLE 10.3

Selected SWAT Model Calibration Parameters Used

SWAT Parameter	St. Joseph	St. Mary's	Peshtigo	Cattaraugus
Alpha_BF	0.02	0.05	0.06	0.20
CN2	–20%	–0.5%	–30%	*
ESCO	*	*	0.01	0.10
GW_Revap	0.20	0.02	0.20	0.20
Manning's coefficient	0.05	0.05	0.05	0.05
REVAPMN	0	*	0	*
SFTMP	1.1	1.1	1.1	1.1
SMTMP	3.5	3.5	4.0	3.0
SURLAG	1.0	0.5	0.5	0.5

* Indicates SWAT Default Value.
Note that only a subset of the SWAT parameters is listed.

TABLES 10.4

The Calibration and Performance Assessment Values for SWAT Model for the Four GLB Watersheds

	Calibration (2000–2005)		Validation (2006–2008)	
	R^2	E	R^2	E
St. Joseph	0.71	0.41	0.76	0.66
St. Mary's	0.83	0.82	0.80	0.79
Peshtigo	0.69	0.43	0.72	0.24
Cattaraugus	0.67	0.57	0.81	0.78

statistics had a wider range of variation (0.41–0.82) than the R^2 values (0.67–0.83). The E values for two watersheds were higher than general threshold value ($E > 0.5$) suggested for the SWAT model calibration (Nair, 2010).

The stream flow outputs from SWAT were validated independently (2006–2008). Table 10.4 shows the statistics for both the calibration and validation periods. A relatively low E value (<0.3) was achieved for the Peshtigo watershed. SWAT model largely overestimated stream flow during the spring months (i.e., April and May). High spatial variability of precipitation during spring months and lack of representative rainfall station might be reasons for the relatively poor model performance (Srinivasan et al., 1998). We developed cross-plots (2006–2008) to compare the predicted monthly stream flow and observed values (Figure 10.2). For all four watersheds, the scatter plots suggested that the SWAT model predicted stream flows matched the USGS observed values reasonably well. The SWAT model performed better for medium flows at monthly intervals. The main problem was the relatively large scattering for the low stream flow values. For the St. Joseph River, St. Mary's, and the Cattaraugus Creek Watersheds, the SWAT model overestimated low flows during the summer months (i.e., July–September). This result was consistent with those reported in similar SWAT model applications. The overestimation of baseflow between rainstorms might contribute to the effects (Van Liew et al., 2007). It should be noted that no detailed agricultural management information was used as SWAT input, but different agricultural practices are likely to affect the water balance (Green et al., 2006).

Sediment Yields

For the selected watersheds, there were no long-term observation data for the calibration of sediment yields. Using the default SWAT parameters, the annual sediment yields (tons/ha/yr) for the St. Mary's watershed (2000–2008) were 2.13. For the same watershed, Whiting (2003) reported the annual sediment yields of 0.60 tons/ha/year. Overestimation of sediment yields has also been reported elsewhere (Chen and Mackay, 2004; Ghidey

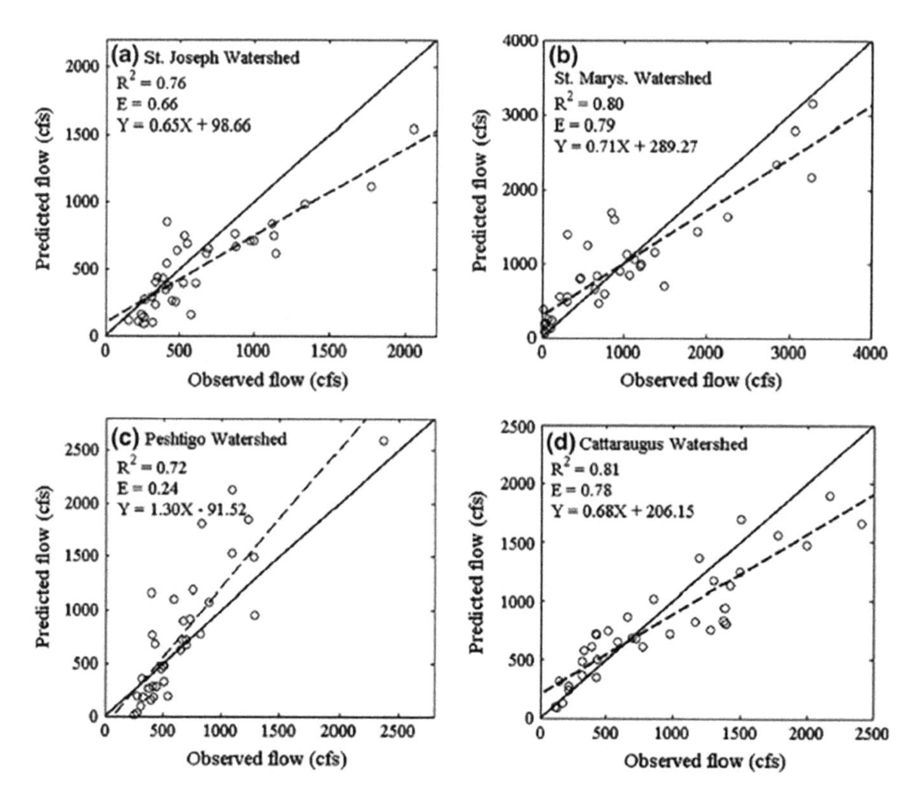

FIGURE 10.2
Comparisons of stream flow estimates from the SWAT model versus the USGS observation data (2006–2008) for (a) St. Joseph River, (b) St. Mary River, (c) Peshtigo River, and (d) Cattaraugus Creek watersheds. Graph ordinates are in predicted flows (cfs) vs abscissa in Observed Flow (cfs).

et al., 2007). One of the main reasons was that SWAT typically uses multiple HRUs for each sub-basin; the sum of the runoff energy from the HRUs did not provide accurate information for transport processes (Chen and Mackay, 2004). Although the hydrology is the driving force of sediment yields, there are many other factors (i.e., support practices (P) factor, slope length factor, slope within HRUs, etc.) that may affect sediment yields. A common practice for the sediment yield calibration is to adjust the USLE_P factor or P factor. Generally, agricultural lands with a slope >5% are terraced (Neitsch et al., 2002). We tested the USLE_P value of 0.5 for the SWAT model (Foster and Highfill, 1983). The slope length factor was also reduced to 30 m (Ambrosio et al., 2007). By adjusting the USLE_P and slope length values, the average annual sediment yield (tons/ha/yr) for the St. Mary's watershed was reduced to 0.58. The newly estimated sediment yields were <3.0% of the reference value. We used the same USLE_P value and slope length factor for all four watersheds due to limited availability of calibration data.

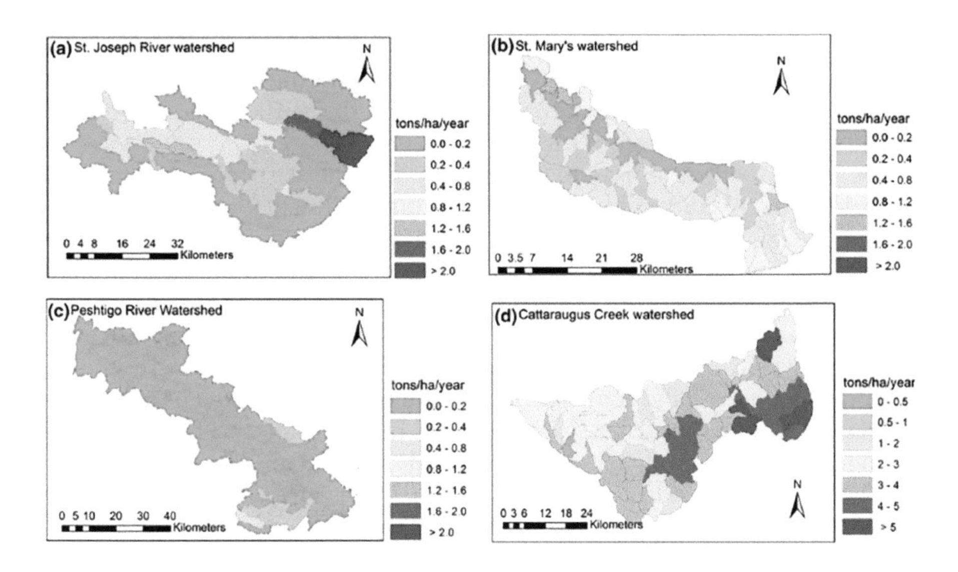

FIGURE 10.3
Average annual sediment yields from the SWAT model (2000–2008) for (a) the St. Joseph River, (b) the St. Mary River, (c) the Peshtigo River, and (d) Cattaraugus Creek watersheds. Each Figure shows sediment yields in tons/ha/year with red, orange and yellow colors (dark grey, medium grey, and light grey, respectively).

The average annual sediment yields (tons/ha/yr) were 0.56, 0.58, 0.12, and 4.44 for the St. Joseph River, St. Mary's, the Peshtigo River, and the Cattaraugus Creek Watersheds, respectively. Figure 10.3 shows the average annual sediment yields at the sub-basin level for four selected watersheds. The Peshtigo River Watershed had the lowest overall sediment yields. A majority of sub-basins generated low sediment yields (<0.12 tons/ha/yr). A few sub-basins located at the southwest area generated slightly higher sediment yields (0.4–0.8 tons/ha/yr). Limited agricultural lands were in these sub-basins. The highest sediment yields were observed for the Cattaraugus Creek Watersheds. The sub-basins with high sediment yields (i.e., >3 tons/ha/year) matched well with the location of agricultural lands, especially areas with relatively high slope values. There were many steep valleys in the Cattaraugus Creek Watershed.

Two future cropland change scenarios were considered. The first scenario was to convert all "other" agricultural row crop types (i.e., sorghum) to corn fields and switch the current/baseline crop rotation into continuous corn. The average annual sediment yields (tons/ha/yr) increased to 0.64, 0.62, 0.17, and 5.97 for the St. Joseph River, St. Mary's, the Peshtigo River, and the Cattaraugus Creek Watersheds, respectively. Compared to the baseline condition, the annual sediment yields increased 14%, 7%, 42%, and 34%. It should be noted that corn and soybean were the dominant row crops for the St. Joseph River and St. Mary's watersheds; thus, the first scenario's impact was mostly due to the shift from corn-soybean rotation to continuous corn. For the Peshtigo River and the Cattaraugus Creek Watersheds, the impacts were mainly due to the shift from corn-alfalfa rotation to continuous corn.

The second scenario was to further expand the corn planting to hay/pasture fields. Accordingly, the average annual sediment yields (tons/ha/yr) for the four watersheds increased to 1.27, 0.77, 0.20, and 7.67. The estimated annual sediment yields increased 33%–127% compared to the baseline conditions. The large increase in sediment yields for St. Joseph watershed and Cattaraugus watershed was attributed to relatively large proportions of hay/pasture lands, which were converted as corn fields in the simulated scenario.

Figure 10.4 shows the comparison of sediment yields for the baseline and the two croplands change scenarios. The increase for the first cropland change scenario (7%–42%) was considered as moderate compared to the second cropland change scenario (33%–127%) because corn and soybeans were already the dominant row crop types in the selected watersheds and the impact was mostly due to the shift from other crop rotations to continuous corn. The large increase of sediment yields for the second cropland

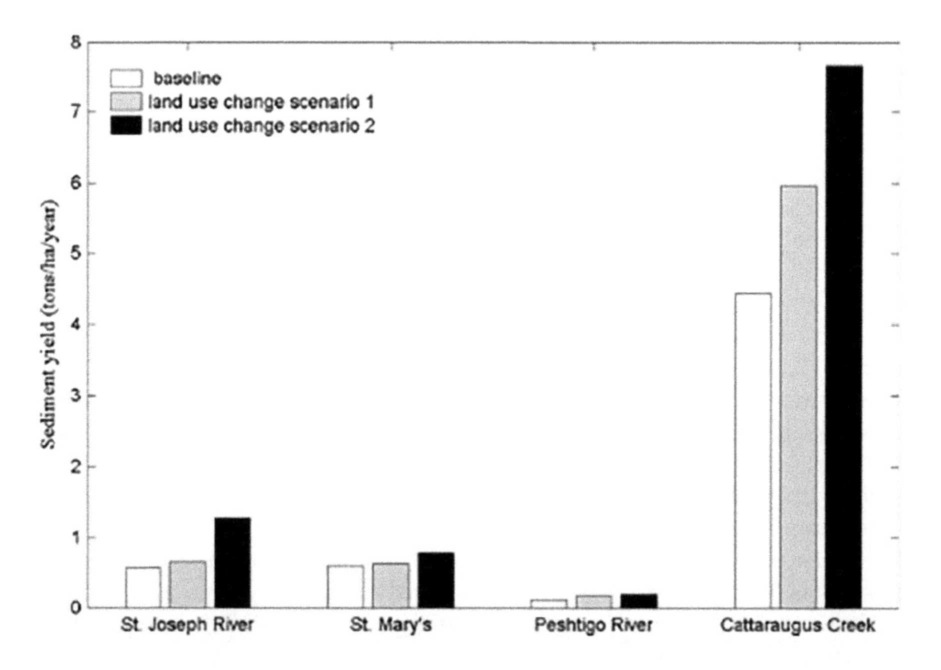

FIGURE 10.4
Comparisons of estimated average annual sediment yields from three land use scenarios. The current baseline condition is included as references (colorless bars). In the first land-use change scenario (scenario 1 in grey), all other agricultural row crop types (i.e., winter wheat) were converted to corn fields. The second scenario was to further expand the corn plantation to hay/pasture field (scenario 2 in black).

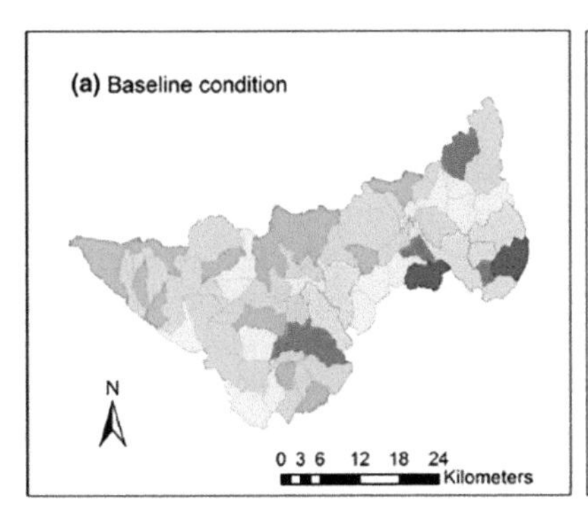

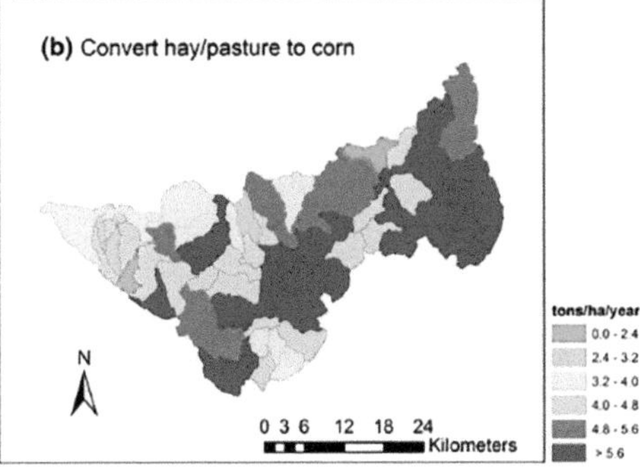

FIGURE 10.5
Average annual sediment yields from the SWAT model for (a) baseline conditions and (b) conversion of hay/pasture to corn field. The subwatersheds are color coded for tons/ha/year with red and orange colors (dark grey and medium grey, respectively) and lower quantity subwatershed in light green, medium green and dark green (grey and light grey, respectively).

change scenario was expected because corn planting generally, generates much higher rates of soil erosion than hay/pasture lands (Claassen et al., 2010). Figure 10.5 compares the spatial distributions of sediment yields at the sub-basin level. Visual interpretation of these two sediment yield maps for the Cattaraugus Creek Watershed suggested that the conversion of hay/pasture to corn planting substantially increased the sediment yields for almost all the sub-basins.

At the period of study, the conversion of pasture/hay field to corn planting was already happening in the Northern Plains (Claassen et al., 2010). In addition to the impacts on the sediment yields, these trends may cause several negative environmental consequences with respect to water quality, biodiversity, and overall ecosystem sustainability (Pimentel and Patzek, 2005). Here the comparison of the sediment yields was conducted using the same climatic conditions for all three land use scenarios. The estimation of sediment yields for future land use change can be complicated by different climate change scenarios. In addition, we only considered two possible agricultural land use change scenarios. Other change scenarios (i.e., urban growth) were not included in the assessment. It should be noted that the historical changes in crop lands in the GLB are much larger than those in the recent decade (USDA-NASS statistics). There is much less land in cropland now than there used to be (i.e., 1930s). The land use change scenarios chosen for this study were modest. More reasonable agricultural land use change scenarios may need to be developed using literature from the economics and policy field, especially from studies that link economic/policy signals and agricultural land use change. In the future study, the SWAT model can be potentially used as an integrated system to assess the coupled climate and land use change impacts on sediment yields and water quality issues.

Conclusions

This study integrated remote sensing-derived products and the Soil and Water Assessment Tool (SWAT) model within a GIS modeling environment to assess the impacts of cropland change on the sediment yield within four selected watersheds in the GLB. The SWAT model was implemented for four selected watersheds in the GLB. We focused on SWAT model calibration and validation for stream flows. For three of the four selected watersheds, the SWAT predicted stream flows matched well with the USGS observation data. For the validation period of 2006–2008, the R^2 values were 0.76, 0.80, 0.72, and 0.81 for the St. Joseph River, St. Mary's, the Peshtigo River, and the Cattaraugus Creek Watersheds, respectively. The SWAT calibration process was further used to estimate sediment yields.

We considered two future agricultural scenarios compared to the current baseline condition, these included the conversion of all "other" row crop types to corn and the conversion of hay/pasture to corn. The average annual sediment yields (tons/ha/yr) for the current baseline condition ranged from 0.12 to 4.44. The average annual sediment yields increased 7%–42% when all "other" row crop types were converted to corn and traditional crop rotations were removed. Further conversion of hay/pasture to corn generated annual sediment yields (tons/ha/yr) of 1.27, 0.77, 0.20, and 7.67 for the St. Joseph River, St. Mary's, the Peshtigo River, and the Cattaraugus Creek Watersheds, respectively. The predicted annual sediment yields increased 33%–127% compared to the baseline conditions.

Acknowledgments

The US Environmental Protection Agency (EPA) funded and partially conducted the research described in this chapter. Although this work was reviewed by EPA and has been approved for publication, it may not necessarily reflect official Agency policy. Mention of any trade names or commercial products does not constitute endorsement or recommendation for use. This research was partially funded by the EPA's Global Earth Observation System of Systems (GEOSS) program under the Advanced Monitoring Initiative (AMI), Grant Number 35.

Part of this contribution is based on "Shao, Y. et al. 2013. Assessing sediment yield for selected watersheds in the Laurentian Great Lakes Basin under future agricultural scenarios. *Environmental Management*, 51, 59–69" by permission from the publisher Springer.

References

Allan, J.D., Erickson, D.L., and Fay, J. 1997. The influence of catchment land use on stream integrity across multiple spatial scales. *Freshwater Biology* 37:149–161.

Ambrosio, J.D., Ward, A., Witter, J., King, K., and Williams, L. 2007. Olentangy River Watershed Total Maximum Daily Load Study. Report to the Environmental Protection Agency.

Arnold, J.G., and Allen, P.M. 1999. Automated methods for estimating baseflow and ground water recharge from streamflow records. *Journal of the American Water Resources Association* 35:411–424.

Arnold, J.G., Srinivasan, R., Muttiah, R.S., and Williams, J.R. 1998. Large area hydrologic modeling and assessment - Part 1: Model development. *Journal of the American Water Resources Association* 34:73–89.

Bhaduri, B., Harbor, J., Engel, B., and Grove, M. 2000. Assessing watershed-scale, long-term hydrologic impacts of land-use change using a GIS-NPS model. *Environmental Management* 26:643–658.

Borah, D.K., and Bera, M. 2004. Watershed-scale hydrologic and nonpoint-source pollution models: Review of applications. *Transactions of the ASAE* 47:789–803.

Chang, J., Hansen, M.C., Pittman, K., Carroll, M., and DiMiceli, C. 2007. Corn and soybean mapping in the United States using MODIS time-series data sets. *Agronomy Journal* 99:1654–1664.

Chen, E., and Mackay, D.S. 2004. Effects of distribution-based parameter aggregation on a spatially distributed agricultural nonpoint source pollution model. *Journal of Hydrology* 295: 211–224.

Claassen, R., Carriazo, F., Cooper, J.C., Hellerstein, D., and Ueda, K. 2010. *Grassland to Cropland Conversion in the Northern Plains: The Role of Crop Insurance, Commodity, and Disaster Programs*. USDA, Washington, DC.

Crosbie B., and Chow-Fraser, P. 1999. Percentage land use in the watershed determines the water and sediment quality of 22 marshes in the Great Lakes basin. Canadian Journal of Fisheries and Aquatic Sciences 56:1781–1791.

Detenbeck, N.E., Galatowitsch, S.M., Atkinson, J., and Ball, H. 1999. Evaluating perturbations and developing restoration strategies for inland wetlands In the Great Lakes Basin. *Wetlands* 19:789–820.

Donner, S.D., and Kucharik, C.J. 2008. Corn-based ethanol production compromises goal of reducing nitrogen export by the Mississippi River. *Proceedings of the National Academy of Sciences of the United States of America* 105(11):4513–4518.

Environment Canada and USEPA, 2003. State of the Great Lakes 2003. Environment Canada and the U.S. Environmental Protection Agency, EPA Report No. 905-R-03-004, 114 pp.

FitzHugh, T.W., and Mackay, D.S. 2000. Impacts of input parameter spatial aggregation on an agricultural nonpoint source pollution model. Journal of Hydrology 236:35–53.

Fohrer, N., Haverkamp, S., Eckhardt, K., and Frede, H.G. 2001. Hydrologic response to land use changes on the catchment scale. *Physics and Chemistry of the Earth Part B-Hydrology Oceans and Atmosphere* 26:577–582.

Foster, G.R., and Highfill, R.E. 1983. Effect of terraces on soil loss: USLE P factor values for terraces. *Journal of Soil and Water Conservation* 38:48–51.

Gassman, P.W. 2008. Simulation assessment of the Boone River watershed: Baseline calibration/validation results and issues, and future research needs. Dissertation, Iowa State University.

Gassman, P.W., Jha, M., Secchi, S., and Arnold, J. 2003. Initial calibration and validation of the SWAT model for the upper Mississippi river basin. In *Proceedings of the 7th International Conference on Diffuse Pollution and Basin Management*, IWA Publisher, Dublin, Ireland, 17–22 August 2003; pp. 10–40.

Gassman, PW., Reyes, M.R., Green, C.H., and Arnold, J.G. 2007. The Soil and Water Assessment Tool: Historical development, applications, and future research directions. *Transactions of the ASABE* 50(4):1211–1250.

Ghidey, F., Sadler, E.J., Lerch, R.N., and Baffaut, C. 2007. Scaling up the SWAT model from Goodwater Creek Experimental Watershed to the Long Branch Watershed. ASABE Paper No. 072043. St. Joseph, MI.

GLC, 2007. The Potential Impacts of Increased Corn Production for Ethanol in the Great Lakes – St. Lawrence River Region. Great Lakes Commission, Ann Arbor, MI. 49p.

Green, C.H., Tomer, M.D., Di, L.M., and Arnold, J.G. 2006. Hydrologic evaluation of the soil and water assessment tool for a large tile-drained watershed in Iowa. *Transactions of the ASABE* 49:413–422.

Harbor, J.M. 1994. A practical method for estimating the impact of land-use change on surface runoff, groundwater recharge and wetland hydrology. *Journal of the American Planning Association* 60:95–108.

Johnson, D.M. 2008. A comparison of coincident Landsat-5 TM and Resourcesat-1 AWiFS imagery for classifying croplands. *Photogrammetric Engineering & Remote Sensing* 74:1413–1423.

Keeney, R., and Hertel, T.W. 2009. The indirect land use impacts of United States biofuel policies: The importance of acreage, yield, and bilateral trade responses. *American Journal of Agricultural Economics* 91:895–909.

Lunetta R.S., Shao Y., Ediriwickrema J., and Lyon J.G. 2010. Monitoring agricultural cropping patterns across the Laurentian Great Lakes Basin using MODIS-NDVI data. *International Journal of Applied Earth Observation and Geoinformation* 12:81–88.

Lunetta, R.S., Greene, R.G., and Lyon, J.G. 2005. Modeling the distribution of diffuse source nitrogen sources and sinks in the Neuse River Basin of North Carolina, USA. *Journal of the American Water Resources Association*, 41(5):1129–1147.

Miller, S.N., Kepner, W.G., Mehaffey, M.H., Hernandez, M., Miller, R.C., Goodrich, D.C., Devonald, K., Heggem, D., and Miller, W.P. 2002. Integrating landscape assessment and hydrologic modeling for land cover change analysis. *Journal of the American Water Resources Association* 38:915–929.

Naef, F., Scherrer, S., and Weiler, M. 2002. A process based assessment of the potential to reduce flood runoff by land use change. *Journal of Hydrology* 267:74–79.

Nair, S.S. 2010. Three Essays on Watershed Modeling, Value of Water Quality and Optimization of Conservation Management. Dissertation, Ohio State University, Columbus, OH.

Neitsch, S.L., Arnold, J.G., Kiniry, J.R., and Williams, J.R. 2002. Soil and Water Assessment Tool User's Manual, version 2000, http://swatmodel.tamu.edu/documentation, last checked December 12, 2021, 781p.

Pimentel, D., and Patzek, T. 2005. Ethanol production using corn, switchgrass, and wood; biodiesel production using soybean and sunflower. *Natural Resources Research* 14:65–76.

Searchinger, T., Heimlich, R., Houghton, R.A., Dong, F.X., Elobeid, A., Fabiosa, J., Tokgoz, S., Hayes, D., and Yu T.H. 2008. Use of US croplands for biofuels increases greenhouse gases through emissions from land-use change. *Science* 319:1238–1240.

Secchi, S., Lyubov, K., Gassman, P.W., and Hart, C. 2011. Land use change in a biofuels hotspot: The case of Iowa, USA. *Biomass and Bioenergy* 35:2391–2400.

Shao, Y., Lunetta, R.S., Ediriwickrema, J., and Liames, J. 2010. Mapping cropland and major crop types across the Great Lakes Basin using MODIS-NDVI Data. *Photogrammetric Engineering & Remote Sensing* 76:73–84.

Srinivasan, R., Ramanarayanan, T.S., Arnold, J.G., and Bednarz, S.T. 1998. Large area hydrologic modeling and assessment - Part II: Model application. *Journal of the American Water Resources Association* 34:91–101.

Statistics Canada, 1998. *National Population Health Survey*. University of Toronto, Data Library, Toronto, Canada.

Stern, A.J., Doraiswamy, P., and Akhmedov, B. 2008. Crop rotation changes in Iowa due to ethanol production. In *International Geoscience and Remote Sensing Symposium, July 6–11, 2008, at Boston, MA*.

Stonefelt, M.D., Fontaine, T.A., and Hotchkiss, R.H. 2000. Impacts of climate change on water yield in the Upper Wind River Basin. *Journal of the American Water Resources Association* 36:321–336.

Tang, Z., Engel, B.A., Pijanowski, B.C., and Lim, K.J. 2005. Forecasting land use change and its environmental impact at a watershed scale. *Journal of Environmental Management* 76:35–45.

Tong, S.T.Y., and Chen, W.L. 2002. Modeling the relationship between land use and surface water quality. *Journal of Environmental Management* 66:377–393.

USEPA, 1997. United States Great Lakes Program Report on the Great Lakes Water Quality Agreement (December) EPA-160-R-97-005. www.epa.gov/glnpo/glwqa/usreport/usreport.pdf

USEPA, 2008. Explore our Multimedia Page. http://www.epa.gov.

Van Liew, M.W., Veith, T.L., Bosch, D., and Arnold, J.G. 2007. Suitability of SWAT for the conservation effects assessment project: A comparison on USDA-ARS experimental watersheds. *Journal of Hydrologic Engineering* 12(2): 173–189.

Ward, A., Trimble, S., Burckhard, S., and Lyon, J. 2015. *Environmental Hydrology*. Third Edition, CRC Press, Boca Raton, FL.

Wardlow, B.D., Egbert, S.L., and Kastens, J.H. 2007. Analysis of time-series MODIS 250 m vegetation index data for crop classification in the US Central Great Plains. *Remote Sensing of Environment* 108:290–310.

Weller, D.E., Jordan, T.E., Correll, D.L., and Liu, Z.J. 2003. Effects of land-use change on nutrient discharges from the Patuxent River watershed. *Estuaries* 26:244–266.

Westcott, P.C., 2007. Ethanol Expansion in the United States: *How will the Agricultural Sector Adjust?* Economic Research Service, United States Department of Agriculture. http://www.ers.usda.gov/publications/fds/2007/05may/fds07d01/

Whiting, P.J. 2003. Estimating TMDL Background Sediment Loading from existing Data – Final Report to the Great Lakes Commission.

Winchell, M., Srinivasan, R., Di, L.M., and Arnold, J.G. 2007. *ArcSWAT Interface for SWAT User's Guide*. Blackland Research Center, Texas Agricultural Experiment station and USDA Agricultural Research Service, Washington, DC.

Wolter, P.T., Johnston, C.A., and Niemi, G.J. 2006. Land use land cover change in the US Great Lakes basin 1992 to 2001. *Journal of Great Lakes Research* 32:607–628.

11

Temporal Downscaling of Daily to Minute Interval Precipitation by Emulator Modeling-Based Genetic Optimization

Venkatesh Budamala, Abhinav Wadhwa, Amit B. Mahindrakar, and B. Srimuruganandam

CONTENTS

Introduction

Multi-temporal downscaling emphasizes the various processes in scientific fields such as hydrological modeling, urban planning, disaster mitigation, water distribution, and so on (Salvadore et al., 2015). Of note, the temporal

DOI: 10.1201/9781003175018-11

downscaling of meteorological variables shows an enormous significance in climate change studies. For instance, within a few minutes, a watershed experiencing intense precipitation gains surface runoff rapidly, potentially causing a flash flood scenario. Hence, it is necessary to analyze and predict the flood vulnerable locations in the watershed. Finer-scale data helps to categorize these flash events as severe or moderate, thereby making it easier to identify the severity extent due to flash floods (Kumar et al., 2018, Ke et al., 2020). However, the acquisition of finer data during these rapid events using traditional means alone becomes a major problem. With the addition of various data sensors including remote sensors, it becomes comparatively easier to obtain the datasets. But still, the data remains limited to daily time scales.

For obtaining finer time-period datasets, it becomes financially infeasible for many parts of the country. Data obtained from sensors or gauging stations provides many dependent anomalies with respect to time, and hence, it becomes complicated to produce the respective future climate data (Ray, 2013, Abatzoglou et al., 2018, Colston et al., 2018, Huntingford et al., 2019). The best alternative to predict and optimize the finer time period is to use Statistical Temporal Downscaling (STD) techniques. Based on the observed time period, the STD identifies the phenomena of historical trends and provides feasible finer time scale data by integrating the incongruities in coarser time scale datasets.

In recent years, regression-based downscaling methods have been favored due to the identification of optimal climate occurrences with minimal complexity and computational burden. The regression-based downscaling employs the transfer function model which relates to the input-output response (Wilby et al., 2002). The transfer function techniques employ statistical linear or nonlinear relationships between observed local variables and General Circulation Model (GCM) or Regional Circulation Model output variables. Instead of GCM outputs, a re-analyzed dataset in which observations and GCMs are combined to generate a synthesized estimate of the state of the system and to fit the regression controls.

STD is essential for both short-term and long-term forecasts, and various researchers have targeted downscale coarser-resolution datasets to finer resolution scales (Mendes and Marengo, 2010, Lee and Jeong, 2014, McIntyre et al., 2016, Schaller et al., 2020, Loganathan and Mahindrakar, 2021, Pan et al., 2021). Recent advances have shown a peculiar development in downscaling of data. Schaller et al. (2020) demonstrated the value and challenges of higher spatial and temporal resolution in simulating flood impacts. Liu et al. (2020) proposed an algorithm for rainfall downscaling to address the problem of insufficient temporal resolution of rainfall during the hydrological simulation. Over different methodologies of temporal downscaling, the regression-based temporal downscaling has demonstrated better approximation power but has been analyzed less in various parts of India (Raje and Mujumdar, 2011, Salvi and Ghosh, 2013, Hernanz et al., 2021).

The generalized procedure for regression-based STD follows the temporal trend of the data and provides an optimal output. An application-based regression technique also reduced the computational load and provides users with an optimum number of options for finer-scale datasets. Specific to Southern parts of India, the effect of finer-scale precipitation helps urban planners to design strategies for reducing the effects of flash floods (Sen, 2010, Srinivasa Rao et al., 2016, Dhiman et al., 2018, National Disaster Management Authority, 2020). To do so, the availability of finer resolution meteorological data is the current need. Hence, the present study focused on developing an adaptive machine learning framework for temporal downscaling one of the meteorological data components, i.e., precipitation. With slight modifications in training datasets, the proposed framework can also be used to downscale the other components for various parts of the country.

Materials and Methods

Framework

The machine learning model represents a relationship between the input and output response of the original simulation model. The efficacy of these models depends on their ability to mimic the function of real-world phenomena. In this study, we developed an Adaptive Emulator Modeling-based Genetic Optimization (AEMGO) framework for the temporal downscaling of precipitation as a variable. The proposed AEMGO framework is shown in Figure 11.1. It consisted of the generation of parameter sets, dimensionality reduction, emulator model fitting, convergence criteria, genetic optimization, and bias correction. The details of the framework are discussed in the later part of this section.

The basic algorithm for the AEMGO framework is as follows:

Step-1: Generation of parameter sets based on the observed data for a meteorological variable at a specific site with fine and coarse time scales (i.e., observed hourly and daily precipitation).

Step-2: Pre-processing of parameter sets: The generated parameter sets may contain a huge discordance and they need to arrange the data with specific consistency. Hence, the feature scaling and dimensionality reduction were applied to the parameter sets which could fit the emulator model with minimal complexity and computational

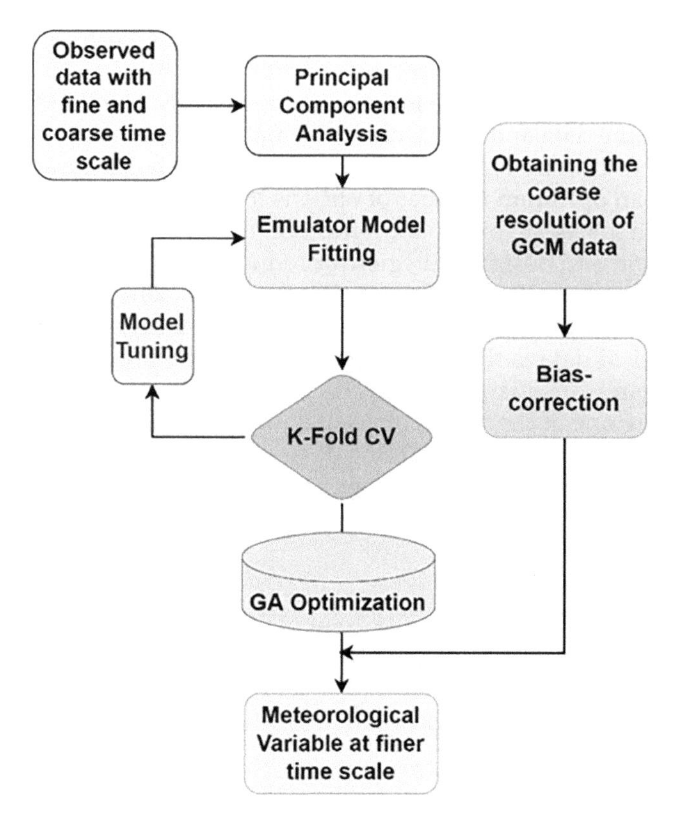

FIGURE 11.1
Stepwise framework for regression-based temporal downscaling of precipitation using AEMGO.

burden. Here, the parameter sets are divided into k-folds and they were transferred to the fitting of the emulator model.

Step-3: Emulator Model Fit: Pre-processing of parameter sets was performed to fit the emulator model for finer and coarse time scales. The accuracy assessment of the emulator model was validated through k-fold cross-validation (CV) as convergence criteria. If criteria were satisfied it directed the fitted model to the genetic optimization phase, otherwise it would tune the model iteratively until the criteria were satisfied.

Step-4: Genetic Algorithm Optimization: Subsequently, with the help of the emulator model, the datasets were analyzed for the optimal finer time scale through Genetic Algorithm (GA) optimization. Here, the GA considered the reproduction, mutation, and crossover for finding the optimal finer time scale based on the coarser time scale. After optimizing the finer time scale ranks obtained from the crossover, it went on to validation. After the satisfactory performance, it was enabled to predict the fine time scale datasets for GCMs.

Step-5: Re-iterations performed to tune the model:

- Obtaining the coarse GCM data of the meteorological variable.
- Bias correction of the coarse time scale GCM data to the coarse observed data for the specific site.
- Temporal downscaling of the bias-corrected GCM data of the meteorological variable.
- Obtaining the future fine time scale data of the meteorological variable.

Emulator Models

Different emulator models have been used for temporal downscaling in this study. Experience with each is illustrative, and hence, they are explained below.

Artificial Neural Networks (ANN)

ANN is becoming more popular for function approximation. The basic structure of ANN comprises hidden neurons, inputs, outputs, and transfer functions. According to the various problems, different methodologies are developed for ANN structures such as pruning or growing strategies, network geometry interpretation, and Bayesian approaches. Apart from these methods, the trial-and-error method is useful for the prediction of many neurons in ANN-based pseudo modeling studies. ANN is comprised of sources of information that are increased by loads and registered by the scientific capacity to decide the enactment work. While another capacity processes the yield of the counterfeit neuron. At that point, it joins artificial neurons to process the data.

Back-propagation is used to prepare neural systems, utilized alongside an improvement routine (like, gradient descent) which expects access to the slope of the loss function for each of the loads in the system. This makes the loads refresh at each iteration, which limits the loss function. Back-propagation registers these slopes deliberately. Back-propagation alongside gradient descent is ostensibly the most essential method for preparing Deep Neural Networks and is known to be the main thrust behind the ongoing rise of deep learning. Any layer of a neural system can be considered as an affine transformation pursued using a non-straight capacity. A vector is considered as an information source duplicated with a grid to deliver a yield, to which a predisposition vector is allotted before going to the outcome through an initiation capacity function, for example, sigmoid.

Classification and Regression Tree (CART)

In machine learning, classification is a two-step process: the learning step and the prediction step. In the learning step, the model is developed based on given training data. In the prediction step, the model is used to predict the response to the given data. A decision tree is one of the easiest and popular classification algorithms to understand and interpret. The decision tree algorithm belongs to the family of supervised learning algorithms. Unlike other supervised learning algorithms, the decision tree algorithm can be used for solving regression and classification problems too. The goal of using a decision tree is to create a training model that is used to predict the class or value of the target by learning simple decision rules inferred from prior data (training data). In decision trees, for predicting a class label for a record, we start from the root of the tree, and we compare the values of the root attributes with the record's attribute. Based on the comparison, we follow the branch corresponding to that value and jump to the next node.

The decision of making strategic splits heavily affects a tree's accuracy. The decision criteria are different for classification and regression trees. Decision trees use multiple algorithms to decide which node is to be split into two or more sub-nodes. The creation of sub-nodes increases the homogeneity of resultant sub-nodes. In other words, we can say that the purity of the node increases with respect to the target variable. The decision tree splits the nodes on all available variables and then selects the split which results in the most homogeneous sub-node network. Hence, the algorithm selection criteria mostly rely on the type of target variables and accuracy range.

K-Nearest Neighbor (K-NN)

K-Nearest Neighbor is one of the simplest machine learning algorithms based on the supervised learning technique. It assumes the similarity between the new case/data and available cases and puts the new case into the category that is most like the available categories. It stores all the available data and classifies a new data point based on the similarity index. This means when new data appears, it can be easily classified into a specified cluster that suits its category by using the K-NN algorithm. The K-NN workings can be explained based on the following algorithm:

Step-1: Select the number of the neighbors (K).

Step-2: Calculate the Euclidean distance of K number of neighbors.

Step-3: Take the K-nearest neighbors as per the calculated Euclidean distance.

Step-4: Among these k neighbors, count the number of the data points in each category.

Step-5: Assign the new data points to that category for which the number of neighbors is maximum. The model is ready for application.

Random Forest (RF)

RF is a popular machine learning algorithm that belongs to the domain of supervised learning techniques. It can be used for both classification and regression problems in machine learning. It is based on the concept of ensemble learning, which is a process of combining multiple classifiers to solve a complex problem and improve the performance of the model. As the name suggests, "Random Forest is a classifier that contains several decision trees on various subsets of the given dataset and takes the average to improve the predictive accuracy of that dataset" (Misra and Li, 2020). Instead of relying on one decision tree, the random forest takes the prediction from each tree based on the majority votes of predictions and predicts the final output. The greater number of trees in the forest leads to higher accuracy and prevents the problem of overfitting.

RF works in two phases: the first is to create the RF by combining the N decision tree, and the second is to make predictions for each tree created in the first phase.

The working process can be explained in the below steps:

- Select random K data points from the training set.
- Build the decision trees associated with the selected data points (subsets).
- Choose the number N for the decision trees that you want to build.

Repeat Steps 1 and 2.

For new data points, find the predictions of each decision tree, and assign the new data points to the category that wins the majority of votes.

Support Vector Machines (SVM)

SVM is designed for both classification and regression problems. Sometimes, it depends on the concept of the basis function (Gaussian kernel) which is used in the Kriging and Radial Basis Function. Additional benefits of SVM over the Kriging and Radial Basis Function is the approximation of variance using the σ-insensitive tube for the formation of support vectors. The benefit of using an σ–insensitive tube in SVM is that it can ignore the errors within a certain distance of true values during the model fitting and can directly control the sensitivities in noise. The foremost advantage of SVM is to formulate the two parameters, i.e., the weight of the regularization term and the radius of the σ-insensitive tube.

Genetic Algorithm Optimization

Evolutionary algorithms (EAs) are populace-based metaheuristic enhancement algorithms that utilize science-enlivened instruments and the existence of the fittest hypothesis to characterize most of the arrangements iteratively. Genetic Algorithms (GAs) are a subclass of evolutionary EAs where the components of the pursuit space are parallel strings or varieties of other basic categories. GAs are PC-based inquiry procedures designed after the genetic systems of natural living beings that have adjusted and prospered by changing in a profoundly focused scenario.

Over a decade, the EAs have evolved with many energizing advances in the utilization of GAs to take care of improvement issues in process control frameworks. GAs are the answer for streamlining complex issues rapidly, reliably, and precisely. As the intricacy of the constant controller expands, the GA applications have developed in more than equivalent measures. The workability of GAs is based on the Darwinian theory of survival of the fittest. GAs may contain a chromosome, a gene, sets of populations, fitness, fitness functions, breeding, mutation, and so forth. The selection of GAs begins with a set of solutions represented by chromosomes, called a population. Solutions from one population are taken and used to form a new population, which is motivated by the probability that the new population will be better than the old one. Further, solutions are selected according to their fitness to form new solutions, profoundly known as offspring. The above process is repeated until some condition is satisfied. With the help of GA optimization, the optimal parameter set of the finer time scale is identified based on the emulator modeling by fitting a coarser time scale data.

Validation Measures and Cross-Validation

In this study, the training of the emulator models was carried out by a *k*-fold CV approach. In this method, the training samples are randomly portioned into *k* subsets among *k* numbered subsets. The training samples are considered such that each kth fold finds the opportunity and *k*-1 would go for training leaving behind one set for testing. The benefit of CV is that it gives a fair prediction of model parameters, and the relative fluctuation is diminished (when contrasted with a split-sample). In this study, the root mean square error was used as a CV score to test the performance of pseudo models (Nash and Sutcliffe, 1970; Viana et al., 2009). The root mean square error gives a global bias measure over the whole design space (Zhang et al., 2018). The computational time is directly proportional to the increase of "*k*" in cross-validation. It was observed that five-fold CV was sufficient to obtain a good performance of the pseudo model and similarly. It has been proven in various hydrologic modeling studies that five-fold was efficient for optimizing the computational burden (Viana et al., 2009; Zhang et al., 2008). Also, a five-fold CV validation was used for the present study as noted in Figure 11.2. The objective function (OF) used in k-fold CV is shown below:

$$\mathrm{OF}_k = \frac{1}{k}\sum_{i=1}^{k} \mathrm{RMSE}_t \tag{11.1}$$

$$\mathrm{RMSE} = \sqrt{\frac{\sum_{i=1}^{n}\left(Y_t - Y_t^1\right)^2}{n}}$$

where n is the number of samples, Y_t is the observed value, and Y_t^1 is the predicted value.

Additionally, three major performance metrics were selected for validating the model outputs: namely, the Nash-Sutcliffe efficiency (NSE; Nash and Sutcliffe, 1970), Percentage of Bias (PBIAS), and the Coefficient of determination (R^2, Jeong et al., 2015). These measures were selected to compare and evaluate the model forecasts under different baselines. The NSE is dimensionless goodness of fit and differentiates the length and thickness of the observed and simulated hydrographs, where its optimum value is 1 (Moriasi et al., 2015). PBIAS tends to show whether the prediction errors are low or high. It provides information such as positive bias and negative bias indicating overestimation and underestimation, respectively. The NSE is one of the best normalized static performance indicators, which determines the relative

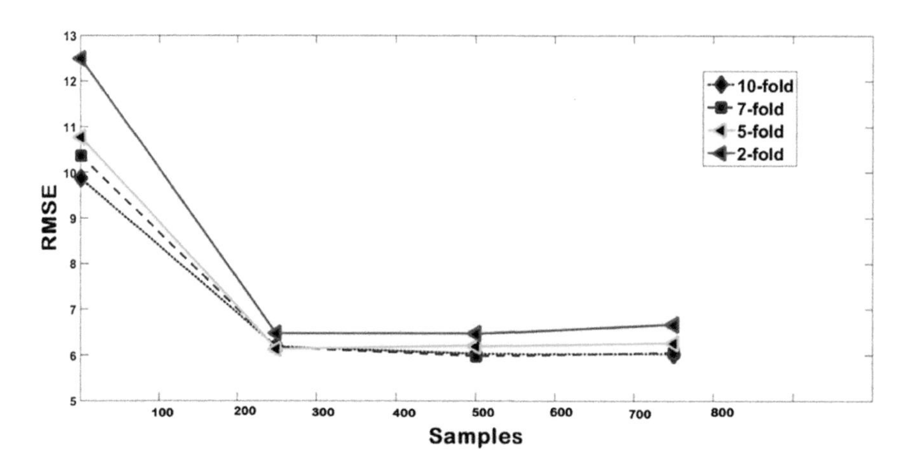

FIGURE 11.2
Comparison of optimal k-fold cross-validation for the present study.

noise to the observed data variance (Nash and Sutcliffe, 1970), and is calculated by:

$$\text{NSE} = 1 - \left[\frac{\sum_{I=1}^{n} \left(Y_i^{\text{OBS}} - Y_i^{\text{SIM}} \right)^2}{\sum_{I=1}^{n} \left(Y_i^{\text{OBS}} - Y_i^{\text{MEAN}} \right)^2} \right] \tag{11.2}$$

where Y_i^{OBS} are the streamflow observed values, Y_i^{SIM} are the simulated or predicted values, Y_i^{MEAN} is the mean of observed values, and n is the number of data points.

The PBIAS measures the mean capability of the model data which tends to be larger or smaller than the observed values. The accurate model simulation with the perfect fit for the PBIAS will be zero. If PBIAS is positive, it indicates underestimation, and when negative, it indicates the overestimation of model output (Moriasi et al., 2015). It is calculated by:

$$\text{PBIAS} = \left[\frac{\sum_{I=1}^{n} \left(Y_i^{\text{OBS}} - Y_i^{\text{SIM}} \right)}{\sum_{I=1}^{n} \left(Y_i^{\text{OBS}} \right)} \right] 100 \tag{11.3}$$

Taking the value of R^2, values near to one declare a 45° best fit line and gave us good predictions. The following formula was used for the evaluation of the correlation among large datasets:

$$R^2 = \frac{n\sum(x_i y_i) - \left(\sum(x_i)\right)\left(\sum(y_i)\right)}{\sqrt{\left[n\sum(x_i)^2 - \left(\sum(x_i)\right)^2\right]\left[n\sum(y_i)^2 - \left(\sum(y_i)\right)^2\right]}} \tag{11.4}$$

where x and y are observed and simulated values, respectively.

TABLE 11.1

Performance Criteria and Classification of Model Performance

Categories	NSE	R^2	PBIAS (%)
Very good	0.75 < NSE ≤ 1.00	0.75 < NSE ≤ 1.00	PBIAS ≤ ±10
Good	0.65 < NSE ≤ 0.75	0.65 < NSE ≤ 0.75	±10 ≤ PBIAS ≤ ±15
Satisfactory	0.50 < NSE ≤ 0.65	0.50 < NSE ≤ 0.65	±15 ≤ PBIAS ≤ ±25
Unsatisfactory	NSE ≤ 0.50	NSE ≤ 0.50	PBIAS ≥ ±25

Based on the NSE, R^2, and PBIAS, the performance of the climate model in the prediction of meteorological variables was classified into four groups as presented in Table 11.1.

According to Moriasi et al. (2015), the best performance criteria of streamflow prediction were considered when NSE > 0.75, R^2 > 0.75, and PBIAS ≤ ± 10. Hence, these threshold limits were considered to evaluate the predictive capabilities of Adaptive Emulator Modeling-based Optimization (AEMO) models.

Study Area and Its Characterization

For this research, daily and 15-minute precipitation data were taken from the Vellore Institute of Technology (VIT), Vellore, weather gauging station (Figure 11.3), which is in the Tamil Nadu state of India (12°58′32″ N, 79°09′41″E), and were employed to present the performance of the suggested model for STD. The study area which falls within the Vellore district has a tropical savanna climate. The temperature in the region varies from 13°C during December and January to 39.1°C during April and May. For validating the framework, the time range of 2019–2020 was considered. In this study, 138 rainy days were used for training and 35 rainy days for testing, and wherein on an average 70% of annual precipitation observed was between the periods of August to November each year.

The climatology of Vellore is influenced by extreme events where it encounters extreme precipitation or generates intense heat that ultimately causes heat waves. Moreover, extreme precipitation often causes urban floods and creates difficulty in assessing stormwater management practices. Hence, the finer time-period precipitation data can provide a mitigation measure for designing urban storm-control management alternatives to enhance the sustainability of the area. Therefore, the VIT station was selected for the current study, and its annual maximum rainfall during different durations was also analyzed for the climate change scenarios. Table 11.2 presents the GCMs datasets which were considered for the temporal downscaling of the VIT station.

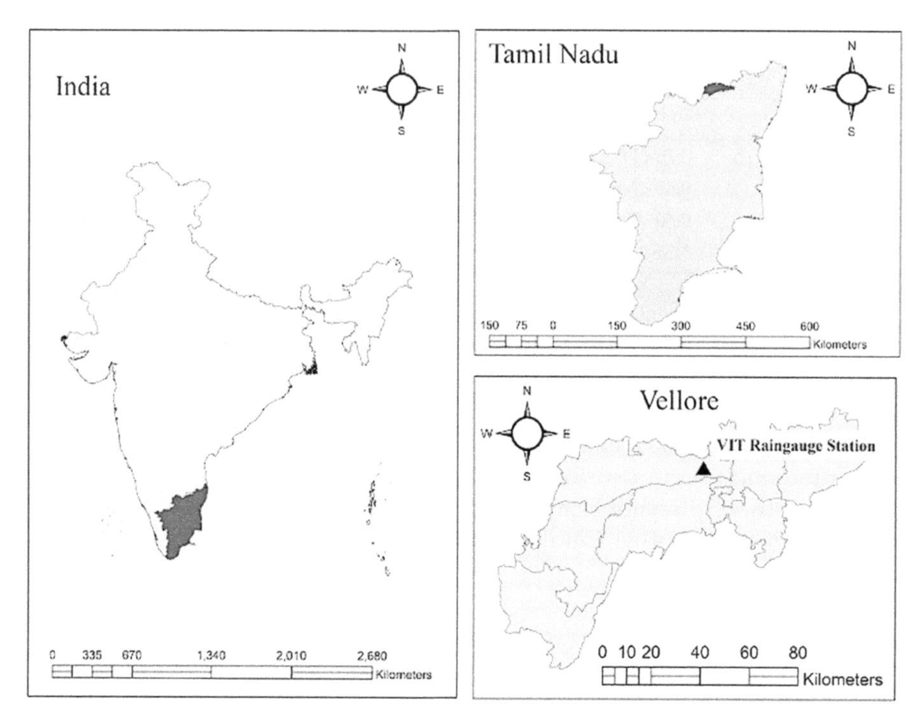

FIGURE 11.3
Map showing the location of the VIT rain gauge station. (a) Map of India, (b) Tamil Nadu, and (c) Velore.

Pre-processing of GCMs

The rainfall data collected for GCMs from different sources provided by the institutes (mentioned in Table 11.2) were converted into daily time-series data. The GCM datasets were then interpolated to station-level data using Bicubic Spline Interpolation techniques to obtain uninterrupted reference data. Usually, GCM datasets from different institutes are at various spatial resolutions which makes it difficult to handle the data. Thus, a technique called re-gridding was used to rescale the entire data to a common (maximum) resolution.

GCM Ranking

In this study, 26 Coupled Model Intercomparison Project (CMIP5)-GCMs were evaluated and ranked using various performance evaluation techniques

TABLE 11.2

Selected 26 CMIP5-GCMs for Evaluation of STD in VIT Station

S.No.	CMIP5 Model ID	Institute and Country of Origin	Atmosphere Horizontal Resolution (°Lat×°Lon)	Atmosphere Eq. Resolution	
				Latitude (km)	Longitude (km)
1	ACCESS-1.0	CSIRO-BOM, Australia	1.9×1.2	210	130
2	ACCESS-1.3	CSIRO-BOM, Australia	1.9×1.2	210	130
3	BCC-CSM1-1-M	BCC, CMA, China	1.1×1.1	120	120
4	BNU-ESM	BNU, China	2.8×2.8	310	310
5	CanCM4	CCCMA, Canada	2.8×2.8	310	310
6	CanESM2	CCCMA, Canada	2.8×2.8	310	310
7	CCSM4	NCAR, USA	1.2×0.9	130	100
8	CMCC-CESM	CMCC, Italy	3.7×3.7	410	410
9	CNRM-CM5	CNRM-CERFACS, France	1.4×1.4	155	155
10	CSIRO-Mk3-6-0	CSIRO-QCCCE, Australia	1.9×1.9	210	210
11	EC-EARTH	EC-EARTH, Europe	1.1×1.1	120	120
12	FGOALS-g2	IAP/LASG, China	2.8×2.8	310	310
13	GFDL-CM3	NOAA, GFDL, USA	2.5×2.0	275	220
14	HadGEM2-AO	NIMR-KMA, Korea	1.9x1.2	210	130
15	HadGEM2-CC	MOHC, UK	1.9×1.2	210	130
16	HadGEM2-ES	MOHC, UK	1.9×1.2	210	130
17	INMCM4	INM, Russia	2.0×1.5	220	165
18	IPSL-CM5A-MR	IPSL, France	2.5×1.3	275	145
19	MIROC5	JAMSTEC, Japan	1.4×1.4	155	155
20	MIROC-ESM	JAMSTEC, Japan	2.8×2.8	310	310
21	MIROC-ESM-CHEM	JAMSTEC, Japan	2.8×2.8	310	310
22	MPI-ESM-LR	MPI-N, Germany	1.9×1.9	210	210
23	MPI-ESM-MR	MPI-N, Germany	1.9×1.9	210	210

(*Continued*)

TABLE 11.2 (*Continued*)

Selected 26 CMIP5-GCMs for Evaluation of STD in VIT Station

S.No.	CMIP5 Model ID	Institute and Country of Origin	Atmosphere Horizontal Resolution (°Lat×°Lon)	Atmosphere Eq. Resolution	
				Latitude (km)	Longitude (km)
24	MRI-CGCM3	MRI, Japan	1.1×1.1	120	120
25	MRI-ESM1	MRI, Japan	1.1×1.1	120	120
26	NorESM1-M	NCC, Norway	2.5×1.9	275	210

ACCESS, The Australian Community Climate and Earth System Simulator; BCC, Beijing Climate Center; BCC-CSM, Beijing Climate Center Climate System Model; CanCM4, The Fourth Generation Coupled Global Climate Model from Canadian Climate Centre; CanESM2, The Second Generation for Earth System Model from Canadian Climate Centre; CCCMA, Canadian Centre for Climate Modelling and Analysis; CCSM, The Community Climate System Model; CESM, The Community Earth System Model; CMCC, Euro-Mediterranean Center on Climate Change; CNRM-CERFACS, National Centre for Meteorological Research and Centre Européen de Recherche et de Formation Avancée en Calcul Scientifique; CSIRO, The Commonwealth Scientific and Industrial Research Organisation; CSIRO-BOM, Commonwealth Scientific and Industrial Research Organisation; Bureau of Meteorology; EC-Earth, A European Community Earth System Model; FGOALS-g2, The flexible global ocean-atmosphere-land system model, Grid-point Version 2; GFDL-CM3, Geophysical Fluid Dynamics Laboratory; HadGEM, Hadley Centre Global Environment Model; IAP/LASG, Institute of Atmospheric Physics, State Key Laboratory of Numerical Modeling for Atmospheric Sciences and Geophysical Fluid Dynamics; INM, Institute for Numerical Mathematics; INMCM, Institute for Numerical Mathematics Climate Model; IPSL-CM5A-MR, Institut Pierre Simon Laplace Model CM5A-MR; JAMSTEC, Japan Agency for Marine-Earth Science and Technology; MIROC5, Model for Interdisciplinary Research on Climate; MOHC, The Met Office Hadley Centre; MPI-ESM, Max-Planck-Institut für Meteorologie; MRI, Meteorological Research Institute; NCAR, National Center for Atmospheric Research; NCC, the Norwegian Climate Centre; NIMR-KMA, National Institute of Meteorological Research (NIMR) and Korea Meteorological Administration (KMA); NOAA, National Oceanic and Atmospheric Administration; NorESM1-M, The Norwegian Earth System Model; QCCCE, Queensland Climate Change Centre of Excellence.

by comparing the historical observed and GCM simulated data. The procedure for evaluating the credibility of CMIP-GCMs in simulating the changes in regional climate was outlined as follows:

- Collection, extraction, and re-gridding of the selected CMIP-GCMs datasets (historical and future scenarios) to a common scale.
- Historical observed weather datasets of the selected station were collected, and missing data were input using the inverse distance weightage interpolation technique.
- The historical and future datasets for selected stations were extracted from the considered GCMs and converted to time series.
- NSE, R2, and PBIAS were calculated for the historical datasets individually for each of 26 GCMs used for evaluating the model skill score.

- Ranks were assigned for each of the mentioned performance evaluation techniques and the final rank for each model was calculated using compromise programming.

Results and Discussion

Selection of Best-Suited Emulator Model

For STD, we evaluated five machine learning or emulator models. Each model has its own characteristics to predict the outputs. So, we ranked all the models based on their complexity (COM), efficiency (EFF), and computational burden (COB) parameters. Initially, EFF projected the optimal results for K-NN, RF, and CART. However, the remaining models performed well but could not meet the threshold limit, even while COM and COB were directly proportional to each other. From the Analytic Hierarchy Process segregation of three emulator models from EFF (K-NN, RF, and CART), the RF was found to require a heavy burden over the remaining models due to the incorporation of many trees and it resulted overfitting. But K-NN and CART consumed less COB in comparison to RF. Finally, CART has better approximation power over the K-NN due to the exploration of the decision tree in terms of both classification and regression concepts. Hence, CART was selected as the emulator model for this study.

Selection of Best-Suited GCMs

The final ranks for the CMIP5-GCMs considered were obtained by multi-criteria ranking with the help of compromise programming. The individual ranks obtained from each of the performance evaluation parameters (NSE, R^2, and PBIAS) were reflected as input criteria for the overall ranking. The results obtained from the credibility ranking for each CMIP5-GCMs are presented in Table 11.3. Evaluation and ranking of the performance of GCMs in simulating the regional climate for the VIT weather gauge station resulted in numerous findings. As shown in Table 11.3, the top three CMIP5 models in simulating the local climatic conditions were CCSM4, MPI-ESM-MR, and CNRM-CM5. The assessment also states that top-performing models in simulating the local climatology over the Vellore were consistent in attaining the hydrological parameters. The comparison of CMIP5-GCMs shows the different capabilities of each model in simulating the regional observed climatic conditions. The performance of each model diverges

TABLE 11.3

Ranking Selected 26 CMIP5 GCMs for Evaluation of STD in VIT Station

	Precipitation			
CMIP5 GCM	**NSE**	R^2	**PBIAS**	**Rank**
ACCESS1-0	22	7	25	24
ACCESS1-3	13	1	14	4
bcc-csm1-1-m	16	8	22	17
BNU-ESM	11	16	21	19
CanCM4	3	12	24	12
CanESM2	14	20	19	22
CCSM4	1	6	2	1
CMCC-CESM	9	9	15	8
CNRM-CM5	6	14	6	3
CSIRO-Mk3-6-0	7	13	13	9
EC-EARTH	4	4	26	10
FGOALS-g2	24	18	8	21
GFDL-CM3	17	2	11	5
HadGEM2-AO	20	24	9	23
HadGEM2-CC	8	19	16	14
HadGEM2-ES	25	25	10	26
inmcm4	19	3	12	11
IPSL-CM5A-MR	23	21	1	15
MIROC-ESM-CHEM	26	15	4	16
MIROC-ESM	5	22	3	6
MIROC5	2	11	17	7
MPI-ESM-LR	18	23	7	20
MPI-ESM-MR	10	10	5	2
MRI-CGCM3	21	5	20	18
MRI-ESM1	15	17	23	13
NorESM1-M	12	26	18	25

ACCESS, The Australian Community Climate and Earth System Simulator; BCC-CSM, Beijing Climate Center Climate System Model; CanCM4, The Fourth Generation Coupled Global Climate Model from Canadian Climate Centre; CanESM2, The Second Generation for Earth System Model from Canadian Climate Centre; CCSM, The Community Climate System Model; CESM, The Community Earth System Model; CMCC, Euro-Mediterranean Center on Climate Change; CNRM, National Centre for Meteorological Research; CSIRO, The Commonwealth Scientific and Industrial Research Organisation; EC-Earth, A European Community Earth System Model; FGOALS-g2, The flexible global ocean-atmosphere-land system model, Grid-point Version 2; GFDL-CM3, Geophysical Fluid Dynamics Laboratory; HadGEM, Hadley Centre Global Environment Model; INMCM, Institute for Numerical Mathematics Climate Model; IPSL-CM5A-MR, Institut Pierre Simon Laplace Model CM5A-MR; MIROC5, Model for Interdisciplinary Research On Climate; MPI-ESM, Max-Planck-Institut für Meteorologie; MRI, Meteorological Research Institute; NorESM1-M, The Norwegian Earth System Model) when considering various seasons, evaluation parameters, and atmospheric parameters. Further, the CCSM4-GCM performed effectively with validating the performance metrics of NSE=0.78, R^2=0.81, and PBIAS=–9.42%, respectively. Hence, the CCSM4-GCM was considered for the temporal downscaling of the VIT weather gauge station with the future emission scenarios of Representative Concentration Pathway (RCP)4.5 and RCP8.5.

Evaluation of RCP Scenarios

After the selection of the best GCMs, the follow-up work was to analyze the climate data for future scenarios (i.e., RCP4.5 and RCP8.5, respectively). Here, RCP4.5 represented the median conditions and RCP8.5 signified the extreme scenarios of varying future climatic conditions. This study analyzed the RCP scenarios of the three best GCMs for a time interval of 15 minutes, 30 minutes, 1 hour, 3 hours, 6 hours, and 12 hours, respectively. The validation of temporal downscaled data was represented through the hydrograph of every 15 minutes intervals and provided in Figure 11.4. Moreover, the 15-minute hydrograph was validated with three different precipitation levels of high peak, medium peak, and low peak levels, respectively. Based on the historical conditions, segregation of the precipitation peak levels was made as 14 mm as high peak, 3 mm as medium peak, and 0.25 mm considered as the low peak. In Figure 11.4, the peak precipitation for different levels was validated with observed data and provided optimal results in every aspect. On the other hand, Figure 11.5 shows the maximum possibility of every 15-minute precipitation interval for both RCP scenarios of 4.5 and 8.5, respectively. From

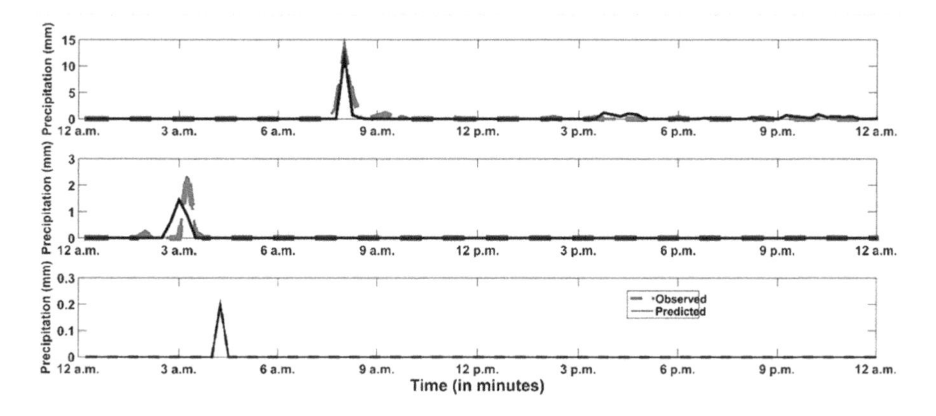

FIGURE 11.4
Validation of temporal downscale data for every 15-minute interval: (a) high-intense event; (b) medium-intense event; and (c) low-intense event.

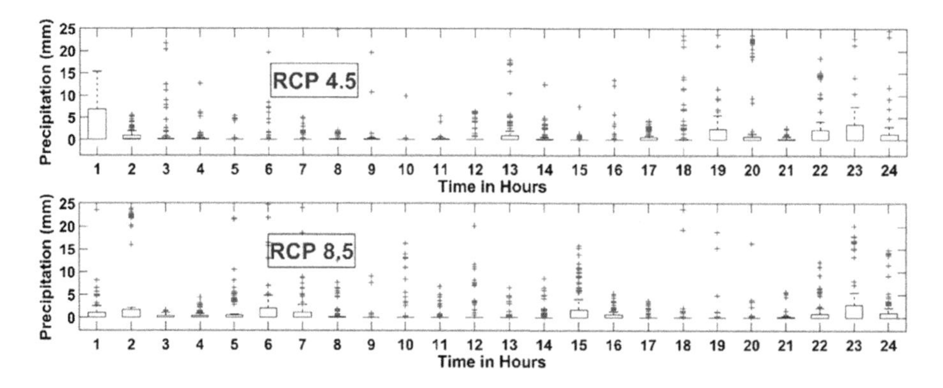

FIGURE 11.5
Future extreme scenarios of precipitation for exemplified CCSM4 climate model data for the most probable rainfall event duration.

the plot, it is projected that the time interval from 12:00 a.m. to 6:00 a.m. contributed heavily over the different intervals for both median and extreme conditions.

Summary

Finer time scale data (i.e., 15 minutes intervals) could serve effectively for different short-term events. Over the decades, acquiring finer data was one of the major problems due to measuring and handling heavy datasets. However, the gauge stations can provide the data at the point of occurrence. Hence, the prediction of future finer time scale data became uncertain. From this drawback, Statistical Temporal Downscaling (STD) came into existence. STD can obtain the relation in historical conditions and predict future scenarios. Moreover, it will bridge the gap between the finer data and future predictions.

This work elaborated on the importance of STD based on regression-based downscaling by comparing the multiple emulator models and GCMS. Additionally, it has provided the selection criteria for GCMs and Emulator models for fitting the temporal downscaling. AEMGO algorithm effectively performed to downscale the coarse to finer time scales by following an adaptive strategy. EMGO was designed based on the pseudo regression approach with genetic optimization. Where the pseudo regression follows the input-output response and replicates the real-world scenario, the Genetic Algorithm (GA) assists in disaggregation from coarser to finer scale by incorporating mutation, crossover, and reproduction concepts. Additionally, the framework enables the adaptive technology to tune the complex parameters for predicting the model outcomes. The proposed framework was applied and validated for the VIT gauge station in Vellore, India. The proposed framework was validated with observed climate data and forecasted the future climatic scenarios by considering the GCMs.

From the reviewed literature, 26 different GCMs were extracted, and a compromise programming-based ranking algorithm was developed. An input-output automated response was identified using a pseudo module proposed in the framework. Further, from a metaheuristic GA-based optimization algorithm, coarser precipitation data was converted to a finer scale. Based on selected climate models, the temporal downscaling of scenarios of RCP4.5 and RCP8.5 were also analyzed. The observation from past daily events in Vellore suggests that in CMIP5 GCM datasets, CCSM4, MPI-ESM-LR, and GFDM-CM3 were the most feasible climate models. Out of all the three above-mentioned climate models, CCSM4 was ranked first in terms of correlation with observed station data. For two GCM emission scenarios of RCP4.5 and RCP8.5, the proposed framework was adopted to obtain the temporal downscaled precipitation data for frequencies (15 minutes, 1 hour, and daily). From the analysis of different rainfall events, it was observed that during rainy days for near-future climate change scenarios, the precipitation will occur mostly during the late-night or early-morning period which thereby resembles the recent scenarios in Vellore.

These framework's efforts were to deliver finer resolution data for an effective understanding of hydrological behavior and to provide an insight into the relationship between gauged and ungauged sub-basins. Moreover, the hydrologic and hydraulic efficiency of any urban flood management alternatives can be improved using the proposed framework. The temporal downscaling of the precipitation output of climate models was critical in the hydrological assessment of climate change, especially for small or urban watersheds. Nonparametric-based modeling of temporal downscaling successfully produced minute-level precipitation conditioned on daily precipitation values. Temporal downscaling of monthly to daily can be a useful tool, but its applications are potentially limited since most of the current climate models produce daily outputs.

References

Abatzoglou, J.T., Dobrowski, S.Z., Parks, S.A. and Hegewisch, K.C. 2018. TerraClimate, a high-resolution global dataset of monthly climate and climatic water balance from 1958–2015, *Sci. Data*, 5(1), 1–12.

Colston, J. M., Ahmed, T., Mahopo, C., and Kang, G. 2018. Evaluating meteorological data from weather stations, and from satellites and global models for a multi-site epidemiological study, *Environ. Res.*, 165, 91–109.

Dhiman, R., VishnuRadhan, R., Eldho, T. I. and Inamdar, A. 2018. Flood risk and adaptation in Indian coastal cities: Recent scenarios, *Appl. Water Sci.*, 9(1), 5.

Errasti, I., Ezcurra, A., Sáenz, J. and Ibarra-Berastegi, G. 2011. Validation of IPCC AR4 models over the Iberian Peninsula, *Theor. Appl. Climatol.*, 103(1–2), 61–79.

Goel, T., Hafkta, R.T. and Shyy, W. 2009. Comparing error estimation measures for polynomial and kriging approximation of noise-free functions, *Struct. Multidiscip. Optim.*, 38(5), 429–442.

Hernanz, A., García-Valero, J.A., Domínguez, M., Ramos-Calzado, P., Pastor-Saavedra, M. A. and Rodríguez-Camino, E. 2021. Evaluation of statistical downscaling methods for climate change projections over Spain, Present conditions with perfect predictors, *Int. J. Climatol.*, 42(2):762–776.

Huntingford, C., Jeffers, E.S., Bonsall, M.B., Christensen, H., Lees, T. and Yang, H. 2019. Machine learning and artificial intelligence to aid climate change research and preparedness, *Environ. Res. Lett.*, 14(12), 124007.

Ke, Q., Tian, X., Bricker, J., Tian, Z., Guan, G., Cai, H., Huang, X., Yang, H. and Liu, J. 2020. Urban pluvial flooding prediction by machine learning approaches – a case study of Shenzhen city, China, *Adv. Water Resour.*, 145, 103719.

Kumar, A., Gupta, A., Bhambri, R., Verma, A., Tiwari, S.K. and Asthana, A. K. L. 2018. Assessment and review of hydrometeorological aspects for cloudburst and flash flood events in the third pole region (Indian Himalaya), *Polar Sci.*, 18, 5–20.

Lee, T. and Jeong, C. 2014. Nonparametric statistical temporal downscaling of daily precipitation to hourly precipitation and implications for climate change scenarios, *J. Hydrol.*, 510,182–196.

Liu, X., Xia, C., Chen, Z. Chai, Y. and Jia, R. 2020. A new framework for rainfall downscaling based on EEMD and an improved fractal interpolation algorithm, *Stoch. Environ. Res. Risk Assess.*, 34(8), 1147–1173.

Loganathan, P. and Mahindrakar, A. B. 2021. Statistical downscaling using principal component regression for climate change impact assessment at the Cauvery river basin, *J. Water Climate Change*, 12(6), 2314–2324.

McIntyre, N., Shi, M. and Onof, C. 2016. Incorporating parameter dependencies into temporal downscaling of extreme rainfall using a random cascade approach, *J. Hydrol.*, 542, 896–912.

Mendes, D. and J. A. Marengo, 2010. Temporal downscaling: A comparison between artificial neural network and autocorrelation techniques over the Amazon Basin in present and future climate change scenarios, *Theor. Appl. Climatol.*, 100(3), 413–421.

Misra, S. and H. Li, 2020. Noninvasive fracture characterization based on the classification of sonic wave travel times, in *Machine Learning for Subsurface Characterization*, Amsterdam, NL: Elsevier, pp. 243–287.

Moriasi, D. N., Gitau, M.W., Pai, N. and Daggupati, P., 2015. Hydrologic and water quality models: Performance measures and evaluation criteria, *Trans. ASABE*, 58(6), 1763–1785.

Nash J. E. and Sutcliffe, J. V., 1970. River flow forecasting through conceptual models part I — A discussion of principles, *J. Hydrol.*, 10(3), 282–290.

National Disaster Management Authority, 2020. National Disaster Management Authority Guidelines - Management of Glacial Lake Outburst Floods (GLOFs). Ministry of Home Affairs - Government of India and Swiss Agency for Development and Cooperation (SDC).

Pan, S., Xu, Y.-P., Gu, H., Bai, Z. and Xuan, W. 2021. Temporary dependency of parameter sensitivity for different flood types, *Hydrology Research*, 52 (5), 990–1014.
Raje, D. and Mujumdar, P.P. 2011. A comparison of three methods for downscaling daily precipitation in the Punjab region, *Hydrol. Process.*, 25(23), 3575–3589.
Ray, B., 2011. *Climate Change: IPCC, Water Crisis, and Policy Riddles with Reference to India and Her Surroundings*, Lanham, MD: Lexington Books.
Salvadore, E., Bronders, J. and Batelaan, O. 2015. Hydrological modelling of urbanized catchments: A review and future directions, *J. Hydrol.*, 529(P1), 62–81.
Salvi, K. S. and Ghosh, S. 2013. High-resolution multisite daily rainfall projections in India with statistical downscaling for climate change impacts assessment, *J. Geophys. Res. Atmos.*, 118(9), 3557–3578.
Schaller, N., Sillmann, J., Muller, M., Haarsma, R., Hazeleger, W., Hegdahl, T., Kelder, T., van den Oord, G., Weerts, A., and Whan, K. 2020. The role of spatial and temporal model resolution in a flood event storyline approach in western Norway, *Weather. Clim. Extremes*, 29, 100259.
Sen, D. 2010. Flood hazards in India and management strategies, in M.K., Jha (Ed.) *Natural and Anthropogenic Disasters*, Dordrecht: Springer Netherlands, pp. 126–146.
Srinivasarao, Ch., Gopinath, K. A., Prasad, J. V. N. S., Prasannakumar, C. and Singh, A.K. 2016. Climate resilient villages for sustainable food security in tropical India: Concept, process, technologies, institutions, and impacts, *Adv. Agron.*, 140, 101–214.
Viana, F. A. C., Haftka, R. T. and Steffen, V. 2009. Multiple surrogates: How cross-validation errors can help us to obtain the best predictor, *Struct. Multidiscip. Optim.*, 39(4), 439–457.
Wilby, R., Dawson, C. and Barrow, E. 2002. SDSM — a decision support tool for the assessment of regional climate change impacts, *Environ. Model. Softw.*, 17(2), 145–157.
Yen, H., Jeong, J., Tseng, W.-H., Kim, M.-K., Records, R. M. and Arabi, M. 2015. Computational procedure for evaluating sampling techniques on watershed model calibration, *J. Hydrol. Eng.*, 20(7) 04014080.
Zhang, Q., Harman, C. J. and Kirchner, J. W. 2018. Evaluation of statistical methods for quantifying fractal scaling in water-quality time series with irregular sampling, *Hydrol. Earth Syst. Sci.*, 22(2), 1175–1192.

Index

D

E

I

J

K

L

M

N

O

P

Q

R

S

T

U

V

W

Z

For Product Safety Concerns and Information please contact our EU representative GPSR@taylorandfrancis.com Taylor & Francis Verlag GmbH, Kaufingerstraße 24, 80331 München, Germany

T - #0170 - 160425 - C59 - 234/156/11 [13] - CB - 9781032006369 - Gloss Lamination